Vallejo artist Cleven "Goodie" Goudea 1932–2015 drew this caricature of Al Zampa and presented it to John Robinson in 2002. *Copyright 2002 C. Goudea, author's collection.*

Bay Area Iron Master
AL ZAMPA
• A LIFE BUILDING BRIDGES •

ISABELLE MAYNARD & JOHN V. ROBINSON | Foreword by Donald Zampa

THE History PRESS

Published by The History Press
Charleston, SC
www.historypress.net

Back cover, inset coin: A commemorative coin produced by Gene Pedrotti and the Bridge Celebration Committee. *Courtesy Gene Pedrotti.*

First published 2015

Manufactured in the United States

ISBN 978.1.46711.913.9

Library of Congress Control Number: 2015952923

To Archie Green (June 29, 1917–March 22, 2009),
who was a good friend and great mentor.

Contents

Foreword

One of my earliest memories of my grandfather was visiting him at his home when I was about nine years old. I walked in to find him in his usual spot, sitting in his secondhand mohair armchair with a Toscanelli cigar in his mouth. He was putting a sticker on his hard hat. He was in his sixties, and as I look back upon that now, I realize that it was the very last project he worked on before hanging up his tool belt and drawing his pension from the Iron Workers Union at the ripe old age of sixty-five.

Years later, I worked with some iron workers who had worked with him on that job, and they told me that he insisted on "connecting" on his last day of work. In the ironworking trade, the men who do the "connecting" are the first to piece together a structure and, therefore, the least protected from falling or being knocked off the iron. They are referred to as a "connector" and are the most confident and nimble people of the trade. So, it was highly unusual for a sixty-five-year-old to be performing such work. But Gramps was always a confident and competitive man. He wanted to prove that he was retiring on his terms, not because he couldn't cut it anymore.

It had been a long journey, some forty-five years, since he had closed up his butcher shop to try his hand at working heavy construction. One of his customers had offered him a shot at working on the new bridge that they were building in his hometown of Crockett. It was a big challenge, but not too big for Alfred Zampa. He was young and stout with a build that earned him the nickname "husky" from his friends. And he was tough—tough and confident enough to fight in a few amateur boxing matches referred to as "smokers."

He turned twenty-one on that first Carquinez Bridge, which opened to the public in May 1927. It was just the beginning of his career of building bridges. He loved his work, even after it almost killed him in 1936. The demands of the trade—to be confident, strong, quick and tough—fit him to a T.

His career as an iron worker changed the course of history for his family. He made a good living, earned a great pension and got his two sons started in the trade. Later on, the respect he earned from his fellow iron workers and his reputation as a "stand-up guy" would play an important role in helping both of his sons get elected into the leadership of the union.

It was my great-grandfather who had left Italy to make a better life in this "land of opportunity," but it was my gramps, Alfred Zampa, who truly

Donald Zampa speaking at the March 2006 retirement ceremony of the 1927 Carquinez Bridge. *Author's collection.*

delivered on that promise of a better life for his family when he became a union iron worker.

Near the end of my grandfather's life, the State of California decided to build another bridge in his hometown. It would be built alongside the two Carquinez spans that he had helped build years earlier: one he started on in the 1920s and the second he worked on in the mid-'50s with his two sons, Dick and Gene, my father and my uncle, respectively.

Gramps attended the groundbreaking for that new span in March 2000 and was introduced as a special guest alongside dignitaries and in front of hundreds of local residents. He would enjoy one last moment in the limelight that day. He passed away just a couple weeks later. He never got to see the bridge finished and never got to hear the news that it would be named for him. The state legislature, under the leadership of state senator John Burton and Governor Gray Davis, declared it the Alfred Zampa Memorial Bridge a few years later.

It is quite an honor to have that bridge bear his name. Even if they had not named it for him, it was quite an honor to have him as my grandfather.

Donald Alfred Zampa
September 2015

Preface

In the summer of 1995, I made the first of many trips down the hill from my Crockett home to Tormey to interview Al Zampa about his life and his thirty-five-year career as a high-steel iron worker as part of an ongoing project for a folklore course at UC Berkeley. We talked in his tiny home, filled with fifty years of bridge memorabilia. Sometimes we talked alone; sometimes Al's grandson Don would sit in. I was lucky that Al Zampa was a friendly and voluble man. He loved to talk about his life and was a natural subject for an oral history. I knew I would write about Al eventually, but I had no specific plans for the material when I collected it.

I did have plans to photograph the construction of the new bridge. So when I learned that the new bridge was to be named in Alfred Zampa's honor the two projects merged into one. There is an old saying that goes, "Chance favors the prepared mind." The confluence of these two events reached fruition in my first book, *Spanning the Strait: Building the Alfred Zampa Memorial Bridge*, in 2003. I was fortunate to have my earlier interviews with Zampa to draw on as I prepared that book for publication. This allowed me to quote Al Zampa extensively and give him a voice in that book.

My interest in Zampa led me to Isabelle Maynard, a writer who, in 1986, interviewed Al Zampa for a play she was writing for the San Francisco Theatre group Tale-Spinners as part of the overall celebration of the fiftieth anniversary of the Golden Gate Bridge. But why was I collecting this material? I wasn't sure. I had the vague feeling that it was important.

To who? Again, I wasn't sure. I just felt that someone should do it and that someone was me.

I always felt *Spanning the Strait* was incomplete in many ways. I thought that Zampa's complete oral history should be published along with Isabelle Maynard's play, *The Ace*. The current book gives me, at long last, the opportunity to present Al Zampa in his own words and publish the play Isabelle Maynard wrote based on the Zampa interviews she collected in 1985 and performed as part of the run-up to the fiftieth anniversary celebration of the opening of the Golden Gate Bridge.

Since those initial interviews, I have continued to gather material about Zampa, the Golden Gate Bridge, the Half Way to Hell Club and other bridges Zampa help build.

Al Zampa was a talkative man. But even a man as naturally gregarious as Zampa doesn't pour forth a chronological narrative in one or two sittings. He was a man who had given over much of his adult life to building bridges, and his work on the Golden Gate Bridge nearly cost him his life. The oral history contained here is the edited product of several interviews conducted from 1986 through 1998. In Part 5, I have included quotes from family to help round out the story of Zampa's life. I hope you like the finished product.

Acknowledgements

In gathering the material presented here I have been helped by many people and organizations that, for many years, have supported my interest in Al Zampa's life story. First among them is Dick Zampa Sr. and the entire Zampa family. I'd also like to thank the Alfred Zampa Memorial Bridge Foundation for its interest in this project. You can learn about the foundation at www.alzampabridge.com.

Along the way, I received valuable help and encouragement from the Contra Costa County Historical Society, the Crockett Historical Society & Museum, the Crockett Library, the Bancroft Library at UC Berkeley, the San Francisco Public Library and the Fund for Labor Culture & History. It was Professor Alan Dundes at UC Berkeley who first suggested I might write about Al Zampa. Archie Green, my friend and mentor, gave me a lot useful advice about how to collect and use the material I was collecting. Thanks to Donald Zampa for first introducing me to his grandfather Al and for writing the foreword to this book. I can't fail to mention Gene Pedrotti, who worked hard for many years to see that the Carquinez Bridges got the media recognition they deserved. He also worked to organize the Alfred Zampa Memorial Bridge celebration. Thanks, Gene!

I'm also pleased to remember my coauthor, Isabelle Maynard (1929–2007), who gave me copies of her Al Zampa interview and her play *The Ace* and encouraged me to try to co-publish the finished product. I'm sorry she did not live to see the finished product. I am delighted, however, to know her play, *The Ace*, will live on between the covers of this book. Thanks go, as well, to my commissioning editor, Megan Laddusaw, at The History Press.

Lastly, I'd like to thank the California State University, East Bay, Office of Faculty Development for a 2015 Faculty in Residence grant that, under the able stewardship of Professor Rita Liberti, gave me the time, money and structure to work on this project.

Were it not for the help of all the above mentioned people, this book might not have been possible.

Thank you, one and all.

Introduction

Born on March 12, 1905, in Selby, California, Alfred Zampa was the first son of Italian immigrants Emilio and Angelina Zampa. Sitting on edge of the San Pablo Bay two miles west of Crockett, Selby was a company town, the ward of the Selby Smelting & Lead Company. Emilio Zampa, like everyone else in town, worked for the smelting company. He worked hard for low pay, and things were often tough for the immigrant family.

The Selby Smelting & Lead Company is gone, and the town is, too. All that's left is the tiny community of Tormey, a small collection of houses that sprang up across Highway 40 from the Selby Smelting Company. Al Zampa settled in Tormey, raised his family in Tormey and retired there, too.

After graduating from high school in Crockett, Zampa, like most young men in town, went to work in the C&H Sugar Refinery. He quickly grew restless working in the sugar factory and quit to become a partner in a short-lived Crockett meat market. He liked to joke that it was the fourth meat market in a town that could support only three.

The first Carquinez Bridge, the longest highway bridge in the world at the time, was special to Zampa because it was a hometown project and because he learned the bridge-building trade there.

Zampa also enjoyed the challenge of the work. He took pride in doing a job that most men couldn't do. The work was hard and dangerous, but it was in his nature to seek challenge. As if bridge building were not dangerous enough, Zampa even boxed for a short time. (He fought under the name "Al Benardo" to hide the fact from his parents.) It didn't take

Ariel view of the American Smelting & Refining Company plant, circa 1950. By the 1970s, environmental concerns and foreign competition made the plant unprofitable, and it was closed. *Courtesy Crockett Historical Museum.*

long for Zampa to figure out that bridge building was something that he could do well.

The first Carquinez Bridge was an architectural marvel that never really got the attention it deserved. But it wasn't for lack of effort. The governors

A large crowd of pedestrians and motorists gather to celebrate the opening of the world's longest highway bridge on May 21, 1927. Earlier that morning, Charles Lindbergh landed in Paris after his transatlantic flight. Lindbergh's flight overshadowed the news of the Carquinez Bridge. *Courtesy Crockett Historical Museum.*

of California, Washington, Oregon and Nevada were present at the first bridge's opening ceremonies. At 2:30 p.m., President Calvin Coolidge, in Washington, D.C., pushed a golden key to signal, via telegraph, the opening of the new span. A large flag was unfurled atop the center tower, the Oakland Municipal Band began to play "The Star Spangled Banner," navy ships from Mare Island fired a volley of shots to salute the new bridge and a flock of carrier pigeons were released to wing the news to Portland, Seattle and Vancouver. The date was May 21, 1927.

A coincidence of history kept the first Carquinez Bridge out of the national spotlight. Earlier that morning, hours before the bridge opened, Charles Lindbergh landed in Paris after completing the first solo transatlantic flight. News of Lindbergh's historic flight quickly upstaged the new bridge. California governor C.C. Young had his speech interrupted several times by newsboys shouting about the Lindbergh landing. "Hell, the newsboys were selling extras about Lindbergh's flight on the bridge that day," Zampa recalled.

After learning the bridge-building trade on the first Carquinez Bridge, Zampa did what many bridge men of his day did: he went on the road building bridges—he "boomed out" in iron worker parlance. Zampa spent

ONE LIFE—

FOR EACH MILLION

One life per million dollars of construction cost was the rule of thumb for bridge builders in the 1930s. *Author's collection.*

the next couple years traveling up California's Central Valley building railroad bridges. In 1928, Zampa married Crockett native Angelina D'amico and settled for a time in Stockton.

In 1933, Zampa returned to the Bay Area, drawn to the challenge of the two great bridges that were just starting. He started work driving the piles for Oakland–San Francisco Bay Bridge. On September 23, 1934, just after the San Francisco Maritime strike, Zampa paid a twenty-five-dollar initiation fee and joined his first union, Piledrivers Local 34.

During building of the Bay Bridge and the Golden Gate Bridge, it was expected that men would die in the process. Among the bridge men, it was said that "the bridge demands life." The superstition among the bridge men was that one life would be lost per million dollars of construction cost. On February 18, 1937, the *San Francisco Chronicle* ran an article titled "One Life—For Each Million" that referred to the ghastly rule of thumb as the "law of concrete and steel."

The first Carquinez Bridge cost $8 million to build, and six men lost their lives during its construction. A total of twenty-four men lost their lives building the Bay Bridge—a fatality rate that chief engineer Charles Purcell apparently found acceptable. Al Zampa was well aware of the ever-present danger during his two-and-a-half-year stint on the Bay Bridge.

In April 1936, Zampa left the Bay Bridge and went to work on the Golden Gate Bridge. Zampa recalled his decision to leave the Bay Bridge: "While I was on the Bay Bridge I could see the towers of the Golden Gate going up. I just knew it was gonna be a great bridge. I wanted to be part of it." The Golden Gate was the first bridge to be built with all-union labor crews, the first to employ safety nets and the first to require that workers wear hard hats. It was at this time that Zampa joined Iron Workers Local 377 in San Francisco.

By the time Zampa got to the Golden Gate Bridge, the towers and the cable spinning were complete, and work was almost ready to begin hanging the steel for the road deck. Zampa went to work in a raising gang as a connector—the men who set the steel beams in place. It was the most perilous job on the bridge. Zampa was on the job about six months when he had the brush with death that would make him famous. The accident occurred on October 19, 1936.

Zampa liked to joke about his near-death experience, but the news story that appeared the next day with the terse caption "Net Grounded; Bridge

The only known photograph of the Half Way to Hell Club, from March 1937. Six of the first ten members pose for news cameras. They are (from left to right) Paul H. Larry, Jack Miller, Edward Stanley, Miles Green, James Roberts and Jack Delaye. Not pictured are Ward Chamberlain, John Perry, George Murray and Al Zampa, who were all still in the hospital. *Author's collection.*

Worker Falls; May Die" hints at the seriousness of the fall. He lay close to dying for three days and was hospitalized for twelve weeks. A bridge man who fell to his death was said to have "gone to hell," so the men who survived their falls into the net decided to call their group the "Half Way to Hell Club." By December 1936, the club had grown to ten men:

Miles Green, age thirty-six
Paul Larry, age thirty-three
James Roberts, age thirty-three
Jack Miller, age thirty
Jack Delaye, age thirty-four
Edward Stanley, age twenty-two
Ward Chamberlain, age twenty-six
George Murray, age thirty-eight
John Perry, age thirty-eight
Al Zampa, age thirty-one

By the time the bridge was finished, the nets had saved, and the club had grown to, nineteen men. A total of eleven men were killed building the Golden Gate Bridge.

The original plan was for the Half Way to Hell Club to meet once a year to commemorate their falls into the nets. But it never really worked out that way. At that time, iron workers were an itinerant breed, following the work from state to state. After the bridge was finished, the men scattered to different jobs throughout the country. But at each important anniversary (ten, twenty-five, fifty years), efforts were made to locate the surviving members, validate their stories and ensure that no interlopers claimed membership in the exclusive club.

No one is really sure what became of the nineteen men who fell into the nets. Photographs are hard to come by. Only one is known for sure: the December 1936 photograph printed on page 21. One reason for this is the name of the club. No news media account from 1936–37 prints the word "hell" in its description of the group. It is referred to as the "Half Way Club" instead. In fact, I have never seen a complete list of names for the Half Way to Hell Club. Whether it ever comes to light, Al Zampa, who lived to age ninety-five, may have been the last surviving member of the club.

Over the years, the fall itself has become the stuff of legend. Various media accounts of the fall have erroneously reported that Zampa fell 740 feet—the height of the towers. But to my knowledge, Zampa himself never claimed a fall of more than 43 feet. During his twelve-week stay in St. Luke's

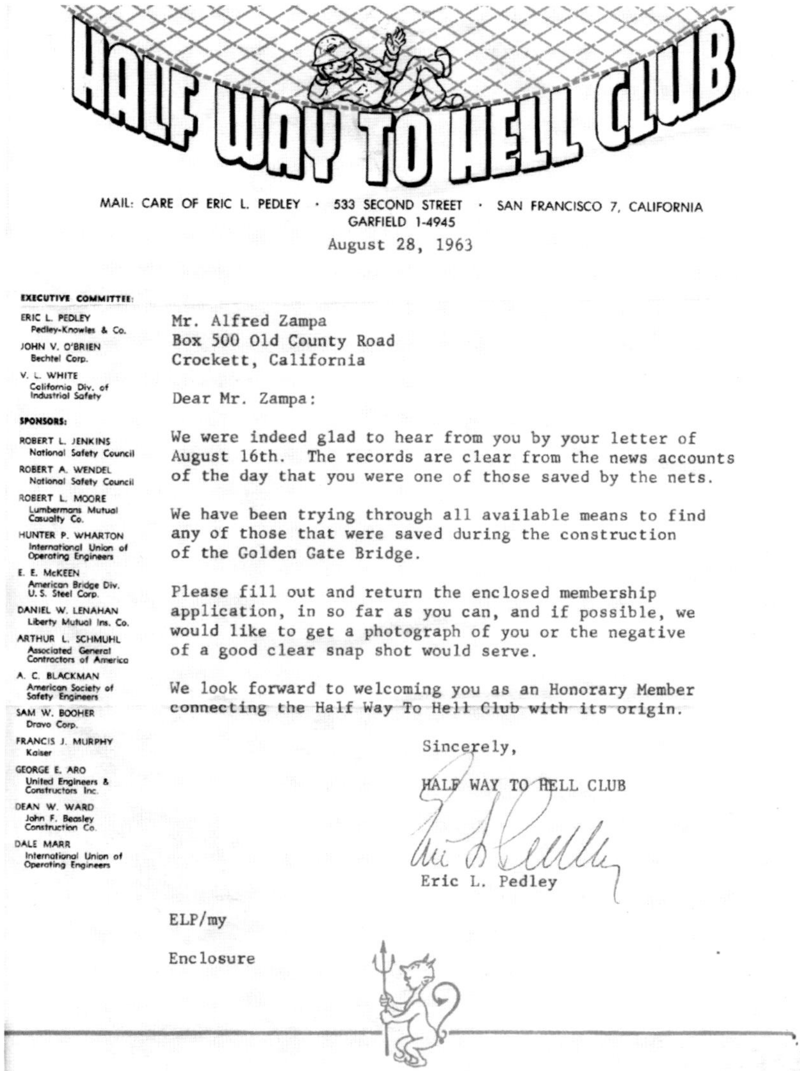

HALF WAY TO HELL CLUB

MAIL: CARE OF ERIC L. PEDLEY · 533 SECOND STREET · SAN FRANCISCO 7, CALIFORNIA
GARFIELD 1-4945

August 28, 1963

EXECUTIVE COMMITTEE:
ERIC L. PEDLEY, Pedley-Knowles & Co.
JOHN V. O'BRIEN, Bechtel Corp.
V. L. WHITE, California Div. of Industrial Safety

SPONSORS:
ROBERT L. JENKINS, National Safety Council
ROBERT A. WENDEL, National Safety Council
ROBERT L. MOORE, Lumbermans Mutual Casualty Co.
HUNTER P. WHARTON, International Union of Operating Engineers
E. E. McKEEN, American Bridge Div. U. S. Steel Corp.
DANIEL W. LENAHAN, Liberty Mutual Ins. Co.
ARTHUR L. SCHMUHL, Associated General Contractors of America
A. C. BLACKMAN, American Society of Safety Engineers
SAM W. BOOHER, Dravo Corp.
FRANCIS J. MURPHY, Kaiser
GEORGE E. ARO, United Engineers & Constructors Inc.
DEAN W. WARD, John F. Beasley Construction Co.
DALE MARR, International Union of Operating Engineers

Mr. Alfred Zampa
Box 500 Old County Road
Crockett, California

Dear Mr. Zampa:

We were indeed glad to hear from you by your letter of August 16th. The records are clear from the news accounts of the day that you were one of those saved by the nets.

We have been trying through all available means to find any of those that were saved during the construction of the Golden Gate Bridge.

Please fill out and return the enclosed membership application, in so far as you can, and if possible, we would like to get a photograph of you or the negative of a good clear snap shot would serve.

We look forward to welcoming you as an Honorary Member connecting the Half Way To Hell Club with its origin.

Sincerely,

HALF WAY TO HELL CLUB

Eric L. Pedley

ELP/my

Enclosure

A 1963 letter from the Zampa family collection hints at early attempts to locate surviving members of the Half Way to Hell Club. *Courtesy Zampa family.*

Hospital, his co-workers from the bridge visited him, convinced that his days as an iron worker were over. To prove to himself and the other bridge men that he still had his confidence, he went out onto the bridge the day he got out of the hospital.

But walking around on the bridge is not the same as working on the bridge. With a brace on his back, he was in no condition to go back to

Three-hundred-foot-tall electrical towers between Crockett and Vallejo were among the projects Zampa tackled upon his return to high ironwork at age thirty-eight. *Photograph by John V. Robinson.*

work. It would be six long years before Zampa returned to work as an iron worker.

In the meantime, Zampa took a job in Vallejo at the Mare Island Naval Shipyard, where he worked as a ship fitter for almost two years. Zampa's decision to work at Mare Island was based on his desire to have a steady job that was close to home so he could spend more time with his two sons, Gene and Dick. Working at the shipyard did not offer the challenge, or the money, that ironwork did, so in July 1942, Zampa reinstated as an iron worker in Oakland Local 378. During this period, he worked for contractors at the C&H Sugar Refinery in Crockett and on the electrical towers that connect Crockett and Vallejo across the Carquinez Strait.

Zampa continued to build bridges, as well. Throughout the 1950s, he worked on the Benicia Bridge, the Richmond–San Rafael Bridge and the second Carquinez Bridge. Working on the second Carquinez Bridge was Al Zampa's greatest source of pride because, on that bridge, he worked alongside

The view looking north shows the 1958 left and 1927 right Carquinez Bridges shortly after the second bridge opened. In many ways, these two spans framed Zampa's working life. *Courtesy Crockett Historical Museum.*

his two sons, Gene and Dick. The second Carquinez Bridge, which cost $26 million and four lives to build, opened for traffic on November 25, 1958.

Zampa was also active in union politics. He served several terms on Local 378's executive board. But when his sons, Dick and Gene, began to assume leadership roles in the union, Al quietly stepped back. Gene and Dick Zampa both served on the executive board in the 1960s. Dick Zampa was elected as the Local 378's president in 1966 and financial secretary in 1968, a position he served in until he was appointed as general organizer and shortly thereafter to general vice-president to the international organization in 1989, a position he retired from in 2008.

Al's grandson Don Zampa is currently the president of the District Council of Iron Workers here in California. Another grandson, Richard Zampa Jr., is the apprentice training coordinator for Local 378. Al Zampa was proud of the fact that three generations of the Zampa family have served as officers of Local 378.

Al Zampa worked as an iron worker until his retirement in 1970 at the age of sixty-five. On his final day on the job, Zampa worked as a connector in a raising gang. (Connecting is one of the most physically demanding jobs in the trade.)

Zampa's early retirement years were largely uneventful. But all of that changed in the mid-1980s, when the Golden Gate's fiftieth anniversary approached and the media's attention naturally turned to the daring men who built the California landmark. Al Zampa and a handful of other surviving bridge men found themselves thrown into the spotlight.

Television reporters began showing up on Zampa's doorstep asking him to comment on the great bridge. One of the first was Charles Kuralt, who profiled Zampa and some of the other old bridge men for a 1983 segment of his CBS show, *On the Road*. The piece was later printed in the 1985 book *On the Road with Charles Kuralt*. Journalists churned out magazine and newspaper articles. Historians and bridge aficionados published books about the bridge and the men who built it. (One of the best of these books is Charles Adams's *Heroes of the Golden Gate*.)

With all the media exposure leading up to the 1987 anniversary celebrations, Al Zampa became the unofficial ambassador to the Bay Bridge and the Golden Gate Bridge. Less widely known was the fact that 1987 was also the sixtieth anniversary for the first Carquinez Bridge.

As part of the general fanfare around the Golden Gate Bridge celebration, San Francisco photographer and award-winning filmmaker Elma Garcia photographed Zampa. The photo shoot produced some stunning images of

Zampa in 1986 with the two Carquinez bridges in the background. *Courtesy of Zampa family.*

A graphic created to publicize the fact that 1987 was also the sixtieth anniversary of the first Carquinez. *Courtesy Crockett Historical Museum.*

In 1987, Elma Garcia made this iconic photograph of eighty-two-year-old Al in his bridge man gear. *Courtesy Zampa family.*

eighty-one-year-old Zampa—steely eyed, lean as a bull rider—in his hard hat and bib overalls, a tool belt slung over his shoulder, looking ready to resume his duties on the bridge that made him famous.

In 1987, Al Zampa's life was given a literary dimension when local writer Isabelle Maynard wrote a play entitled *The Ace*. The play, which carried the subtitle "An Iron Worker's Story of Heroism, Risk, and Recognition on the Golden Gate Bridge," was based on Al Zampa's experience on the bridge.

In August 2001, I drove to Emeryville, California, to meet with Isabelle Maynard and discuss Al Zampa, *The Ace* and the writing process. Maynard rewarded my visit with drafts of the script, photographs of Zampa at the theater and numerous tales about meeting Al and Angelina Zampa.

The play was performed at the Life on the Water Theater at Fort Mason in San Francisco from May 13 to June 7, 1987. *The Ace* was well received by audiences and got generally good reviews. It was later adapted and performed on KPFA radio in Berkeley. Al Zampa and his family were

Bay Area artist Doug Minkler created this Half Way to Hell Club poster to accompany the play *The Ace*. *Author's collection.*

present for the opening night performance of *The Ace*. After the play, Zampa was introduced to the audience and, along with Isabelle Maynard, answered questions about his experiences on the bridge.

To supplement the play, the lobby of the theater was decorated with pro-union posters. One of the posters, created by Berkeley graphic artist Doug Minkler, celebrated the Half Way to Hell Club, as well as acknowledging the significant fact that the job was the first major construction project in the Bay Area that was a 100 percent union job.

During that jubilee year, the retired bridge man was invited to numerous celebrations surrounding the three bridges and received numerous proclamations, congratulatory letters and general encomiums from civic organizations and politicians ranging from San Francisco mayor Dianne Feinstein to California governor George Deukmejian.

Proud, but never vain, Zampa did occasionally express regret that the men who built the great bridges didn't get recognized sooner for their achievement. There is, after all, near the toll plaza at the Golden Gate Bridge, a section of main cable and a bronze statue of bridge designer Joseph Strauss. Why is there no statue honoring the riveters who made Strauss's dream a reality?

In 2000, local artist Vince Ramos memorialized Al in the mural depicting Al and the Golden Gate Bridge on a retaining wall near Zampa's home in Tormey. *Photograph by John V. Robinson.*

The groundbreaking ceremony for the new Carquinez Bridge was on March 3, 2000. Al Zampa was invited to speak. It was his last public appearance. It's fitting that it was at the site of a new bridge.

On April 23, 2000, just a few weeks after the groundbreaking for the new span, Alfred Zampa quietly passed away at age ninety-five. It was Easter Sunday. His passing was front-page news in the local papers. It was as if the beginning of the new bridge was the culmination of his life as a bridge builder. Not long after Al Zampa's passing, a local man, Vince Ramos, memorialized the ace by painting a large mural on a retaining wall overlooking the old house in Tormey where the old bridge man had lived for many years.

Later that same year, in August 2000, California state senator John Burton introduced Senate Resolution 97 to have the third Carquinez Bridge officially named in honor of Alfred Zampa. That resolution was passed and signed into law on September 12, 2000.

Built in his hometown, by iron workers from his home local, the new bridge is a fitting tribute to Alfred Zampa and the generation of men who built the early bridges, as well as a testament to the strength and courage of all bridge builders past, present and future.

Part 1

Tough Times in a Small Town

I was born here in Selby, California, and raised here, too. Selby was a collection of houses connected to the Selby Smelting & Lead Company, which later became American Smelting & Refining Company. Most of the town is gone. What's left is all part of Crockett now. All the little company towns got absorbed into Crockett over the years. I went to John Swett High School in Crockett. When I got out of high school we moved into Crockett. This house I live in now is in Tormey. It was owned originally by the smelting company, too. It was practically a village. These dozen or so houses here in Tormey is all that's left of the little villages that belonged to the smelting company. See, all the other houses are gone. About 1970 they closed the refinery, tore down the company houses, and moved east. They got a place in Missouri now, I hear.

We had a nice little town with a post office, train depot, hotel and company store. The company would give my dad scripts to buy stuff in the store, you know, when we was short of money, so we could buy everything: clothes, groceries, et cetera. The company would take it out of his paycheck. Well, that little store has a little bit of everything, you know. Wasn't too much around then, but the company was real good to the employees. We had company baseball teams, and when the first movies came out, they'd show us a movie every week. We had a local Boy Scout troop, and in Crockett, the company would sponsor events for the boys. They'd give us a dollar or two to get the boys lunch. It was a good company, and they moved away. Anyway, that was a lead refinery, and lead is common with gold, silver, copper—everything. They

This little island of homes, sandwiched between an oil refinery and the San Pablo Bay, is all that remains of the communities of Tormey and Selby. Zampa lived his retirement years here. *Author's collection.*

The western half of Tormey, pictured here in 1950, was owned by the smelting company. The refinery closed, and the homes were boarded up and finally razed in 1973. *Courtesy Crockett Historical Museum.*

didn't produce the ore. The big mining companies would ship in the raw ore in long strings of railroad cars. The Selby Plant would smelt the ore and separate the lead from the gold and other metals. Now, the government wanted the gold and silver to make into coins. Coins used to be made from gold and silver. I guess that plant made a lot of money over the years.

The only trouble was a lot of the fellows used to get sick; they'd get lead poisoning. We called it "leaded" in them days. My dad, Emilio, he worked seven days a week; of course, he had to—he had five kids. Those days there was no welfare, no nothing. If he didn't work, we'd starve.

Well, he ended up being a fireman at the plant. He'd fire all the furnaces at the plant—to process the ore. Like I said, it's a good town. When we had to go to school in Crockett, we went on a train. It was only a mile or so, but this was before cars got big. So people took trains. There was a train depot in every village: Rodeo, Oleum, Tormey, Selby and Crockett, where John Swett High School was. It'd leave here at 7:15 a.m., and we get up to school you know (two miles away), we'd get there about 7:20 or 7:25. School didn't start until 8:30 so we went to the Crockett Club. They had pool tables, had a library, and all the newspapers. When school let out we took the train back. At the end of the year (according to your attendance) you'd get that money back that you spent for the train.

Anyway, guys who worked at the smelting plant got leaded from breathing the fumes. It's full of lead gas. They'd get sick and go to the hospital; maybe they'd be there for three or four weeks. They had to drink a lot of milk—that kind of helped it—but it was rough. Later on, they cleaned up the plant quite a bit. In the early days, it was bad. People didn't die from the lead outright. They'd recover, but they always had lead in them. Later on, a lot of them workers started dying, you know, early deaths, but they could never pinpoint if it was the lead that killed them.

My dad got "leaded" and went in the hospital. My mother said it was the only vacation he ever got. When he died, I thought it was the lead that probably killed him. No one could prove nothing. No, in them days, they didn't do anything about suing. Now, every time you turn around, somebody is suing somebody for something, which isn't right. My dad was always angry about the conditions at the plant—not safe, you know? But everyone just took it as a fact of life. They had to. People had families to take care of—what else could they do? They sacrificed a lot to give us an opportunity for a better life. I remember the way it was here. If somebody was doing some work on a house or digging a ditch or something, people might walk by, make a little small talk, then pitch in and help. You never asked anybody to do anything

California & Hawaiian Sugar Company, pictured here in 1921, was the chief employer of the area for many years. C&H is still in business and celebrated its 100th anniversary in 2006. *Courtesy Crockett Historical Museum.*

for you, they'd just come and do it. Then some day you would do it for them. It was a pretty good life. Money was tight. People got more things now, but there's less community and family values. When I was a kid we didn't have enough to eat sometimes. My mom would make bread for the week. It would get stale but not get moldy. So for breakfast we'd have a piece of hard bread soaked in wine. We lived on bread, cheese, salami, wine and whatever vegetables were in season from our garden.

A lot of guys wanted to work at the smelting plant or the C&H [California & Hawaiian] Sugar Refinery. They didn't want to go to high school. The only thing that attracted them to go to John Swett was sports. Guys wanted to play baseball, basketball or run track. We didn't have football then. They had cut out football when I was there in the 1920s. Everybody was getting hurt, so they had to cut it out. We didn't have the equipment or the training that the kids have now. Guys would wear a thick sweater, some old sweat pants and an old leather helmet. So I never got to play football in high school.

In 1927, they opened the new John Swett High School. That's the school everyone today knows. Well in 1927, it was modern. They had everything: a beautiful auditorium, a gymnasium, shop classes to teach machine and woodwork to give working-class kids some skills. Guys who came out of that shop program had no trouble finding work at the brickyard in Port Costa or the smelting plant or one of the big oil refineries near here. But most went to the C&H Sugar Refinery. That was a good company, too; it took care of its people, too. It had parks, a swimming pool and other activities for the people of town. At Christmastime, it'd take all the kids and give them all Christmas presents, gift bags, too, with candy, an orange and little toys. It sponsored dances for the people. It was real nice. Everything was free for the employees and [residents of] the surrounding town.

Generally, the people owned their houses, but the land belonged to the company. We paid a small fee to lease the land. Now, that was true in a lot of places around here. People leased land from the railroads, too. They'd build houses or garages. We had free water, but we didn't have no lights and no gas. We had coal oil lamps and a wood stove. We had to heat our own water on the stove. He got permission to raise rabbits or grow vegetables or whatever. At one time, we had four horses.

My dad always obeyed the law, even during Prohibition, when guys would make extra money selling wine. He never even bootlegged. A lot of the others guys were bootleggers in them days—making it to sell, I mean. You see, our folks were Italian. During Prohibition, the government let us make 250 gallons of wine per family—but nobody else could do it. We had it on the others; they'd come around because they wanted to get some of that wine. And some guys would sell it, too, but not my dad. So, we had it on them there. See, you'd go to work, [and] they wouldn't let anybody bring a bottle of wine to work other than the Italian or Spanish. My dad could take wine with him, with his lunch, because he had a dry lunch (salami, cheese and a piece of bread—a poor workingman's lunch). They'd let the Italians bring wine into the plant, but if they catch somebody else with a bottle, they'd fire them. C&H Sugar was that way, too. Now, I don't know if it was official policy or just informal. When I was a kid my dad had to work every day, seven days a week. There was five of us, three boys and two girls. If he didn't work, we'd starve. We had nothing.

We got our first car in 1923. It was a Nash. We were one of the first ones to have a car, too. I had two brothers, Albert and Freddie. Albert was only fourteen months younger than me. We would fight about who was going to get the car when we was in high school. It's funny now that I think of it, but my dad never drove; my mom never drove. Me and my brother were the ones who used to drive. We had a lot of fun when it was my turn to have it. Sometimes I would get it when it was his turn. We had a lot of arguments and fights. We had different friends. My friends were a little on the wild side. Freddie was more conservative. I just had more guts, I guess. I'd do anything I wanted. The only thing that I was afraid of was my dad. He tried to keep us straight.

Really, the only thing I did that Mom and Dad wouldn't approve of was boxing. I was a club fighter for a few months when I was sixteen or seventeen. I lost my first fight, and then I won fourteen in a row. I fought under the name Al Benardo so my dad wouldn't find out. When I was young, I felt sorry for the girls. They couldn't do nothing. I mean nothing! Back then a respectable young woman had to go out with a chaperone. I used to have to take my sister to a dance or else she couldn't go.

My dad was strict. He was from the old country, so he had Old World values. We didn't get into much trouble, just little things that would come up at school. If we caused any big trouble he woulda beat the hell out of us. We weren't afraid of the cops, my brother and me—I wasn't anyhow. The only thing that held me back was my dad and mom and what they would think of me. I didn't want to do anything that reflected bad on my family.

My mom and dad used to talk broken English, you know, half English, half Italian. My dad was a good talker. I guess that's where I got the gift of gab. But my mom, she didn't talk good English. In the evening they hung around their own people. She didn't know how to say some things. She'd have to point to this and point to that in the local stores. It was kind of tough. We were Italians, and some of the locals didn't want us here. Just like they do the blacks and the Mexicans now. It was kind of hard for us. It didn't bother me so much. Then, later on, I understood; I didn't understand what it was all about at the time when I was young, you know. I know what my dad had to go through that I didn't. "Damn dagos, taking all the jobs," things like that.

I was terrible myself sometimes. I'd start talking pig Latin to my folks. The old man wanted to shoot me. I felt awful about it later. You know, it's a funny thing, but it's true. I thought my mom and dad were stupid until I was a junior in high school. I thought I was the smart one. Then I started to realize what they done for me. I was ashamed of them sometimes. Honest I was. But I guess all teenagers go through that stage. You see, young people are bright, but they don't have any life experience. So they don't have wisdom. It takes smarts and life experience to have wisdom. Anyway that's how it seems after raising my kids, and now I see it in my grandkids. When I got to be a junior and senior in high school, I thought to myself, "Gee, I can't understand how my dad and mom got to be so smart in the last two years."

So, my childhood was good, I'd say. Times were tough, but we lived in a close-knit community. So even the prejudice we experienced—being called "dumb dago" and such things—didn't really affect us too much. We didn't have much money, but we had some good times. My dad died in 1947; he was still telling me what I should do when I was thirty years old. That's how much I grew to respect what he tried to do for me.

When I was nineteen, my dad and uncle set me up with a partner in the meat market business in Crockett. See, when I was a teenager, I worked in a grocery store after school and on Saturdays. One half was a meat market and the other half was groceries. So I thought I knew about the business. It was something I felt I could do. (Boy, was I wrong.) In those days we used to deliver. We did pretty good at first, but we gave out too much credit. Soon we had about $4,000 outstanding on the books (and that's a lot of money in 1924), and we paid cash for the meat. People would run a $30 or $40 bill. Sometimes they would give us $10 on it; sometimes they wouldn't pay anything. Hell, you can't stay in business long like that. We were the fourth meat market in a town that could support only three. Finally, I said, "I want out. This is too slow for me."

By this time, 1925, they had started building the first Carquinez Bridge. A superintendent for Missouri Valley Bridge & Iron happened to rent the downstairs of my dad's house. He was a customer at my meat market. I told him to settle his bill because I was going to quit and just close the place up. He said, "Well, Al, if you don't have nothing else to do, why don't you come out and work on the bridge?" But I didn't go straight to the bridge. My dad didn't like the idea of bridge work. So, I bumped around for a few months. I worked for C&H Sugar. I worked at Union Oil for a few weeks. I even drove a team of four mules for a few weeks. It wasn't for me. I couldn't go for that. I didn't want to work in a factory. My dad was worried when I couldn't make a go of it at these local jobs. So, I thought about it, and I said, hell, why not. I was open for a challenge, and the bridge was a big challenge.

Part 2

A Job for Desperate Men

Bridge men are a breed to themselves, like a migratory bird with an uncanny ability to sense the next big job. They will show up from all over the country at the site of a new bridge.
—Joseph Strauss

That first Carquinez Bridge was a private enterprise. Two local businessmen named Aven Hanford and Oscar Klatt dreamed it up. They wanted a faster way to get their goods across the water between Contra Costa and Solano Counties. So they formed the American Toll Bridge Company, sold stock and started to build the bridge. I'm sure it was a complicated job, but in hindsight it came together pretty fast. On the Carquinez Bridge, some guys lived in Crockett and some guys lived in Vallejo. I lived and worked on the Crockett side. Crockett was little Reno, you could play anything you wanted—you could play craps, you could play roulette, cards. They had women, too. There was this place on the edge of town, near the bridge, that had about eighteen girls. It was run by a woman called Big Marie and a guy known as Harry the Pimp. For $1.50, you could have any girl you wanted. I guess some guys had them all at one time or another.

There used to be eight or ten restaurants in Crockett then—bars, dance halls, a movie house. It's dead now. The bridge men were making big money. I was twenty years old, and I was making fifteen dollars a day in 1926. My dad was making four dollars a day at the Selby smelting plant. See, we'd work eight hours—that's ten dollars—and then two hours at

The men in the foreground work on the footings at Crockett. Big Marie's bar and brothel is in the background. *Courtesy Crockett Historical Museum.*

double time, which paid me fifteen bucks. That went on for four months, ten hours a day, five days a week. Guys were used to working for two or three dollars a day. I thought I had it made. But I pissed the money away like most iron workers do.

I turned twenty-one on the first Carquinez Bridge. It was the hardest bridge for me because I had to do a day's work while I was learning the bridge-building trade. I came close to falling a couple times. Sometimes I was scared, which is good, it keeps you sharp. We had no safety lines, no hard hats, no nets, nothing. We had a saying back then: "One hand for the work and one hand for yourself." I've heard sailors had a similar saying: "One hand for the ship, one hand for yourself."

I first went to work on the foundations, the footings and the piers for Missouri Valley Bridge & Iron. On the ground, digging out the muck putting the piles in place and then the rebar for the concrete. After we finished the piers, American Bridge came to put up the steel. American Bridge is the erection division of United States Steel. They put up both Carquinez Bridges, one in 1927 and one in 1958.

One of the American Bridge guys asked me if I wanted to work up in the air, and I looked up at those guys and thought, "If they can do it, I can do it." And so I went to work for American Bridge. That was in 1926. Now, I'd never been higher up than a garage roof before I went out to the bridge. I've always said it takes 90 percent guts and 10 percent know-how to build a bridge. And you cannot fake guts.

I worked with a lot of different fellows. Some of them had no schooling at all. They could hardly read or write. Everything was "dees, dem and dose [these, them and those]." I wasn't gonna correct their grammar; they'd think I was a lawyer and beat the hell out of me. Anyway, you learn bridge building by doing it. I was fast. I was a quick learner. So I got a reputation as a good hand. If you're good, they move you around. So over the course of two years, I learned to drive piles, drive rivets, work signaling the cranes and set up the hoisting engines. See, the 1920s was interesting because cars were getting big but we needed roads and bridges to take advantage of them. When that bridge started, they used steam shovels and horse-drawn wagons to haul the dirt away. When that bridge opened, it was a new age. It was the age of automobiles and airplanes.

Everybody wanted me to work for them. Sometimes I'd take a job, and if I didn't like the job or the men, I'd quit. I knew I could get another job. One of the things I liked about the work was the freedom. Every time you go to a new job, you meet new men from all walks of life. Some you like; some you don't like.

Another thing about that bridge, it was designed by David Steinman. It was his first big project. He went on to great fame. He designed the Mackinac Bridge in Michigan. Steinman was the first engineer to put earthquake dampeners on a bridge. I helped put them in. They looked

The south suspended span floated into position in March 1927. *Courtesy Crockett Historical Museum.*

like big shock absorbers you might see on a modern car. I guess they worked well enough. The bridge stood up to that big quake in 1989 [Loma Prieta, 7.0].

Mr. Steinman also came up with the idea of using counterweights to lift those two big suspended spans into place. That was something to see. They built the two suspended spans on piers west of the bridge. Then they loaded them onto barges and floated each one out into place beneath the bridge. Now, above, they had four large boxes filled with wet sand that weighed enough to counterbalance the weight of the suspended spans. When everything was in place, they released the sand boxes, and as they went down to the barges, the spans rose up into place. I don't think that has ever been done before. There were thousands of local people on both shores to watch that show.

That first Carquinez was the first big bridge in the Bay Area. That bridge was open ten years before the Golden Gate Bridge. I don't think Steinman got the credit he deserved for that bridge. And I don't think that first bridge got the recognition it deserved. On the day it opened, there was thousands of people to see it. Governors from three states were there making speeches. Then something extraordinary happened. Word started circulating among the crowd that Charles Lindbergh had landed in Paris, the first man to fly nonstop from America to Europe.

This photograph from 1929 shows Al Zampa piloting the tugboat *Frank C.* Zampa liked to remind people that old-time bridge men did everything on the job site. *Courtesy Zampa family.*

This photograph from 2015 shows the San Joaquin River Bridge that Zampa worked on in 1930. *Courtesy Zampa family.*

It completely disrupted the events that day. Hell, newsboys were selling extras about Lindbergh's flight on the bridge that very day. That was May 21, 1927.

After I finished up on the Carquinez Bridge, I went down to Oceanside, again for Missouri Valley Bridge & Iron. I'd be on each job for seven or eight months until we got done and then we'd move on to the next bridge. They were small railroad bridges but made of heavy iron. Sometimes they'd be drawbridges. I did whatever I could do: I drove pile, I worked on derrick barges, I was a rigger, I drove a tugboat, hell I even painted a few of them. Back then, before the unions got big, we were called "bridge men" because we did it all. I learned bridge work from the ground up.

About 1929, I was called to Stockton to build a bridge across the San Joaquin River for Santa Fe Railroad. That one was a bascule bridge, which is a drawbridge with a large counterweight to balance the upward swing of the bridge deck. That bridge is still in service.

I got married while I was in Stockton. I married Angelina. I was twenty-three years old. I knew her in grammar school. Her folks and my folks came from the same town in Italy. It's a little town like this [one]. They came over here, and they knew one another. Then I married her. We're the only ones who stayed true (an Italian and an Italian). She was born here, too—born and raised here like I was. She was married before we met to an older guy, an arranged marriage. She was not happy, but neither was he, I gather. So when we met and fell

While working near Stockton, Al married Angelina D'amico, pictured here in 1925. They stayed married for fifty years. *Courtesy Zampa family.*

in love, she decided to leave her first husband. That was not an easy thing to do. People did not do that back then. We lived in a small town and there was scandal and gossip. But there is a lot of goodwill in small towns like Crockett, and we got through it.

Over the course of two years, I built three bridges across the Old River, the Middle River and the San Joaquin River. All near Stockton. I got married to Angelina there. She came out to be with me, and we got married and lived in Stockton. My first son, Gene, was born in Stockton.

From Stockton, we went to Arizona, and we took the baby with us (he was only about a year old then). I worked in Arizona for about seven or eight months. I got done with that in 1931, and from there, they sent us out to Amarillo, Texas—forty miles out of Amarillo, way out in the sticks. I'd usually be one of the first there. I'd look around to see what I needed. See, we bring a few key men and hire the rest local. That's usually in the contract. On this job, the guys would come out by train. They'd stop the train at the job site, forty miles outside Amarillo. They were willing to work for fifteen cents an hour. Dollar and a half a day for ten hours' work. They came by the trainload. We had to put a notice in the paper saying we couldn't use anybody. We had got enough.

The bridge company built a camp there. Married men built little houses if they were going to say a long time. I lived in a tent (me and Angelina). I put in a ten by twelve-foot board floor. It had a four-lid wood stove, and I made a bed in a corner for baby Gene. Everything we had in a ten by twelve cabin-tent (me, her, and the baby). We lived on the site. We ate camp chow. We carried our own water. Angelina was terrific. How many women today would move to a construction camp and live in a tent? Every now and then, bandits would show up, pull a big fucking hog-leg [pistol] and rob the commissary, rob where the guys sleep. Maybe get $100 or $150. They never caught them bastards. All that said, no one ever bothered us. We was lucky in that way.

Well this was an unusual job for me. This was when I worked as a sand hog. Just after we got there, it rained real hard one night, and they had a flood that just wiped out the work we'd done on the bridge. I think it's on the Canadian River. So we had to wait for the material to come down from Oregon. Hell, that took almost four weeks to get there.

Anyway, the super said, "Al, we got no work outside until the piling comes down from Oregon. If you want to go sand-hogging, I'll send you down there. That work will keep going." So I said I might as well try it. I knew all about it by talking to the other guys.

Zampa at work unloading timber from a train near Amarillo, Texas, circa 1931. *Courtesy Zampa family.*

Sand hogs are like miners, only they just dig tunnels, like for a subway. They also work in underwater chambers called "caissons" using compressed air to hold the water out. On the Texas job, I worked ninety-two feet below the water, at forty-two pounds per square inch of air pressure. Normal air pressure is about fourteen pounds per square inch. For every twenty-six inches deeper you go, you need an extra pound of air pressure to keep out the water.

See, before, it was all blacks doing it because it was hard, dangerous work. But eventually, white guys started doing it, too. It was quite an experience. I guess it was about fifty feet deep already (now that's underneath the surface). They place the caisson on the bottom and guys go down inside the chamber and start digging. The pier has got a steel cutting edge all around, and you got wood on the sides there (six by twelves). That's the planks, you know? And they're big like railroad ties. You'd have to work in there, dig underneath—underneath the concrete and underneath the pier. Now the pressure holds the water out, and to dig it, we'd have to have air spades and cut up the mud and gravel until we reached bedrock. The air pressure keeps the water out where they're cutting away the mud and muck. It won't go down, so we'd dig it out. Like I said, the air wants to escape, and we had these big hoses. You fed it a little water, a little mud with a target (like a two by twelve) with a handle on it and keep digging so much at a time. Get all the mud, rocks, gravel, everything, and it goes up through the pier. It shoots it out like a cannon outside because of the difference in pressure. So, sometimes we got to ditch this far underneath, and it wouldn't go down. So it helps to let the air out. Let the pressure down, let out. You blow it out, and the whole pier comes down like a hydrologic jack. It just comes down a few inches at a time. But man it was scary. You'd think the whole pier was coming on you with the water rushing in. You got boots on, but you got to keep walking—moving your feet; otherwise, when the air comes back, they'll have to dig you out. Anyhow, the limit is 110 feet. That's fifty-two pounds per square inch.

So, anyhow, when you climb down through a manhole with an airlock on the top and the bottom. You climb down to the second hatch, and they raise the pressure in the chamber. When that air equalized in that lock, the bottom door drops. Then you go down the ladder and go to work. See, your body can take being pressurized fast but can't be decompressed as quickly. That has to be done slowly, or you'll get sick. We called that caisson's disease.

As we cleared out the muck, more weight would be added on top. The caisson would sink down a foot or so, and we'd dig out more muck. The pier would be right over the top of us. We were underneath the water, underneath the pier. That pier comes down on you. There is a tremendous weight on top of the work chamber. Sometimes we'd hit some rock, and we'd have to shoot dynamite down in there to break up that rock. We'd light the fuse and hurry up that shaft before it went off. Well, if we'd had a blowout and the air gone through there, then we'd drown like rats. That's when I found out what my father meant when he said it was a "job for desperate men."

I think I might be the only guy alive who did both jobs. I worked six hundred feet up on the towers of the Golden Gate (changing the sheaves on the derrick atop the towers), and I've worked ninety feet below ground under compressed air. I was up to heaven and down to hell. Well, after that Texas job finished up, I was supposed to go to Henderson, Kentucky, but I never got there. My brother Freddie sent me a telegram that my mom was sick and might be dying. She wished I would come back and see her before she goes, so we came back to Crockett. I stayed until Mom got well. That was 1931, and I worked in California mostly after that. Gene was getting to be school age. I figured that I didn't want him to go to this school and that school and

This 1931 photograph shows Zampa (from left, third man kneeling) and his crew on a job in Texas. The rest of the men are unidentified. *Courtesy Zampa family.*

live a camp life. So I came back to Crockett, near where I was born in Selby, and rented us a house. I wanted him to go to my school where I went. In fact, when I graduated, there was only five of us in that eighth-grade class. I thought my kids was it; I just loved my kids—they were everything to me and Angelina.

Part 3

Half Way to Hell

I've climbed halfway to heaven and fallen halfway to hell. Neither place wanted me so I just kept on working.
—Al Zampa

I started on the Bay Bridge in early 1933 working for Missouri Valley Bridge & Iron, putting in the piles for the foundations on the Oakland side. I didn't have to belong to the union to work on the Bay Bridge. Unions weren't big yet in California. The Bay Bridge was 60/40—mostly non-union. Most of the guys in my gang were in the union. I saw the writing on the wall. I saw the unions were going to be big. So I joined Local 34 of the Pile Drivers.

The Bay Bridge started first, before the Golden Gate. A lot of different companies had the contracts to build the different parts of the Bay Bridge. As I said, I worked for Missouri Valley on the piers, the foundations, as a pile butt [pile driver]. I worked on the deepest pier in the world, E-3. It was 242 feet deep. It was an open pier: they build a box with steel slip forms and sink them on the bottom, pump out the water and the guys climb down in there to work. That's called a cofferdam. Some water leaks in, but they keep pumping out. When they dug down deep enough, they'd put in the rebar and then pour the concrete. They put the concrete down there with two-yard buckets.

When they poured the concrete, they had guys down in there with shovels to spread it around. The big buckets couldn't get it in the corners. They'd have four or five guys at each corner to shovel the concrete into place. Building a bridge still requires a lot of hand labor. All that mud from the piers and the dirt from the tunnel through Yerba Buena Island was dumped in the Bay to create Treasure Island, which was used for the 1939 Golden Gate International Exposition.

You probably heard about people getting buried alive in the concrete, but that's all bullshit. The concrete is hard, not soft and watery like you might think. They have the least water they can in the concrete because it's setting in the Bay. I walked on it. It's just like walking on the ground. It's just sand, gravel, water and cement to bind it all together. But most big jobs have a legend about some poor guy falling into the concrete and getting buried alive. I heard it about Hoover Dam, for example, but it's just not true. Nowadays the pile drivers do all the foundation work. So when I started I did everything. I mean I had ten years' experience working on bridges. All bridges are a little different. But [there is] nothing like the Oakland Bay Bridge. That bridge is every kind of bridge rolled into one: the trusses, the five camelback spans, the cantilever bridge that leads up to the tunnel through Yerba Buena Island, then two

Zampa, in 1986, on Yerba Buena Island with the cantilever span of the Bay Bridge in the background. *Courtesy Zampa family.*

This photograph from June 1935 shows a crew of iron workers at the center anchorage early in the cable spinning operations. Al Zampa is second from right. *Courtesy Bancroft Library.*

suspension spans end to end. The Golden Gate is just one type—it's a suspension bridge.

In 1935, I was working on the suspension spans of the Bay Bridge. Once on the footbridge of the suspension spans, when we was spinning the main cables, I saw a guy get hit in the mouth with a length of heavy chain. He staggered back and then down he went. He hit the water flat and just floated there. All his clothes were torn and fanned out, just like an open coat. We lost twenty-four men on the Bay Bridge. Whenever somebody got killed, we went home for the day. Guys go get drunk. It makes you think; it might be you next time. When you lose a guy, you really think about what you're doing. A lot of guys think they can do this work, but anyone might go up and have a look around. But it's not enough just to go up, you got to go to work when you get there. And you gotta work fast.

On the Bay Bridge we could tie off when we were up top and on point. But when you're putting up the steel, you can't tie off because you have got to be able to move. Once you get on station, the riveters and the bolter-ups, they're tied off. They have a float to work off. A "float" is a makeshift scaffold you hang from the steel. It's a four- by six-foot panel of boards with four one-inch-thick rope lines in each corner to hold it up. Sometimes it's the only place to stand.

By the summer of 1935, I was working on the San Francisco side of the Bay Bridge. Now, from Yerba Buena Island to San Francisco is not one long bridge. It is two suspension bridges laid end to end, with a massive center anchorage in the middle of it. I started on the suspension spans when they were spinning the cables. When that was finished, I started putting up the steel for the road deck. Those double-deck sections were built down on Islais Creek, not far from where Candlestick Park is now. They were loaded onto a barge and floated out to us. We raised them into place using gantry cranes. So when they started raising the steel, the raising gang had to climb down the towers to hook on and bring up the steel a couple of times a day. Now, inside those towers are hollow, dozens of cells, ladders and hatches. You could get lost in there. Guys did get lost in there. I said, "Bullshit to that. I'm gonna drive rivets."

Trusses in place on the suspension portion of the Bay Bridge. *Author's collection.*

A rivet gang consisted of four men: A cooker heated the rivets in a small forge that (looked like a barbecue) 'til it was red hot. Then, he grabbed it with tongs and tossed it to the catcher, who caught the heated rivet in a large tin cup and placed the red-hot rivet in the hole. The buck-up man held the rivet fast with a large hand tool called a "horse-cock," and the gun-man used a pneumatic rivet gun to pound the free end of the rivet to a round head.

Since I was small I could get into tight places and into the box girders. So I would buck up the rivets. The engineers come out, and they check everything. If you are in the riveting gang, they check all the rivets. They have to be perfect. Before we move the floats and the forge, one of the inspectors goes out, puts a half dollar on the rivet head and taps it with a hammer to be sure it's not loose. If any are bad, we bust them out and replace them. We had an air-powered chisel called a "hell dog" to bust that rivet out. Imagine a jackhammer in the street. Now use that to bust the head off a rivet—in the air. They're big and heavy. If you are not careful, the plunger will come out, guts and all. Then you gotta climb down and get that damn plunger. Do that twice, and you're gone. Someone else has got your job. We were supposed to drive 250 rivets a day. A good gang, same four guys who work together all the time, could get 300 or more. We could get 300 on most days.

Bridge work is hard. A bridge man has more talent. He has to have more fortitude. Working over the water is tougher since you have to deal with wind more. On a high-rise, they erect two floors at a time. There are big wooden planks on the derrick floor. When two floors of iron are in place, they jump the derrick to the top, put down planks and start putting up the next two floors. If you fall, you only fall two floors. Unless you're on the outside [perimeter]. When you're on the outside making a connection, or driving rivets, well, then you fall all the way to the street, "go in the hole" is how we iron workers say it. But, like I said, it's harder to work on the bridge. You've got to be a little better. I don't know what it is, but you've got to have confidence and a lot of nerve. You got to know you can do it. A guy can't work scared. You've got to be relaxed. Just getting up there is a job, but then you've got to go to work. If you can't do it, you can't stay. On the Bay Bridge, to get a job on the steel as an iron worker, you had to go see the foreman—on top of one of the towers. The foreman would see you coming. The guys who scampered right up he had a good feeling for. It was like an interview in itself. Nowadays, if a guy is good they say that he's "bridgy"—like good enough to work on a bridge.

When we were setting the steel on the Bay Bridge, we could see them spinning the cable on the Golden Gate. So there was a sense of competition among the men, even though the Bay Bridge was a much bigger project overall than the Golden Gate. Most of the news stories were about Strauss and the Golden Gate Bridge. We could see the Golden Gate Bridge was going to be great. It was just impressive.

So in the late summer of 1936, I went over to the Golden Gate Bridge. I knew it was gonna be historic, and I wanted to be part of that. That was the first 100 percent union job. It had to be; otherwise, it wouldn't have worked. We had iron workers come from all over the country to work on the bridge. If they weren't in the union and didn't join, they couldn't get a slip to go to work. As far as I know I'm one of the few native Californians who ever worked on the bridge. Now they had some sort of residency requirement to work on the bridge. But there were rooming houses in San Francisco where the owners would write a letter saying you had lived there a year or so and you could go to work. That's why in the news article about my fall I'm listed with a San Francisco address. That was a rooming house.

Those bridges were built by iron workers from all over the United States: New York, Pennsylvania, Florida and Oklahoma. The good ones all came out. But a guy wouldn't come all that way if he didn't know he was good. We call the good ones "aces." I was an "ace." I thought I was pretty good until I saw some of them. An ace is a fellow who can do the work right, you know, all day up in the air, and he does the work quick. No scaffold, no nothing. Just guts. A lot of them were as good as me or better. But over time, I lost track of them. One by one, I'd get news that they fell or lost a finger in the rigging or broke a leg or plain get too old and broke down to do the work anymore.

On the Golden Gate every once in a while some guy would have enough; maybe he had a close call, so he would quit. Or some guy gets hurt, so there's room for one more. Then they'd go down and pick a few out. We'd be up on the deck setting iron, and I'd look down and see a hundred or more guys at the gates. They were out of work carpenters, iron workers and laborers, and they were cooking bacon butts and beans in a five-gallon tin over an open fire, like a hobo camp. So one of the superintendents would go down to the gate and ask for their experience. Are they in the union? They didn't even have to write out an application. Somebody almost always knew them, you know. They'd say, "I know this fellow here; he's OK." Or, "This guy worked on such a job. He's a good man." A "good hand" is what we called them. That's how they got hired. You could be the best hand, and if no one vouched for you, well, you're out. It was the Depression, so the best men from all over the country came out. Things got sorted out pretty quick. How a guy

Al Zampa, as he appeared in 1935 in full bridge man's regalia. *Courtesy Zampa family.*

handled himself those first five minutes in the air would tell you all you need to know about a guy. The one thing we all had in common was they, like me, wanted to work on the Golden Gate Bridge—the most famous bridge there is.

Yeah, you get a wonderful feel being up there—proud. You hope for the best. You say "well, I wonder if this thing will hold up? I hope those engineers are right." God bless them, they were. The only time I was scared was when we was going to the job and coming back. We used to go from the San Francisco side, they would ferry us over on a little passenger boat, and those big waves coming at us. On the crew boat was the only time I was leery. But some of the guys would shoot craps on the deck of the crew boat on the way back in the afternoon. But I wasn't leery up on the bridge.

To unwind, we'd go over to Sausalito, you know, to raise a little hell. You see, you worked five days a week, so Saturdays and Sundays were ours to recuperate. Oh, it was wild—nice. I had a lot of good times there. I would drink mostly. Most bridge men drink like hell anyway. We used to drink and shoot dice and play poker; sometimes, we'd be there all day. Whenever we got rained out, we'd head over to a couple of iron worker bars that the guys knew about. Some guys had girls they'd meet up with them. We called them "rainy-day girls." But that's another story.

We'd get only a half hour for lunch, so we'd bring a brown bag up there. When we were up in the air, we would just bring it up in our overalls and set it someplace and eat it up there and then go back to work. It was a thrill every time—lunch was a thrill. Sit with a group of guys and talk about the last big job we was on. Complain about the way things are done on this job. Guess what? They did it better on the last job. They say it's a soldier's right to complain. Well, iron workers, too. I loved it. Ha, you had to love it or you couldn't do the work.

The wind was a thrill a minute. You've got to lean into the wind. But when you go near the towers, they were so wide, there's no wind there. But when you got past the towers, well, the wind would get at you again. It was so windy that we had little wires, it was number nine–gauge wire [about a quarter inch in diameter] that we tied on the suspender cables for a makeshift handrail. Once in a while we could grab it. The top of the cords was about two and a half feet wide. But still it was slick and almost nothing to grab on to. The wind was strong moving from one place to another. And the weather was different on different parts of the bridge. We'd be at the base of the tower off-loading a barge, and it would be cold and foggy. We'd get on the construction elevator to go up top and pass through the fog into bright sunshine. The tower tops were above the layer of fog.

But rain didn't bother us too much. If we got too wet, we would knock off; we'd go home for the day or head over to Sausalito. It was too dangerous to work in the rain. Bethlehem Steel and American Bridge were self-insured, so if they had an accident, they'd have to pay. That was one of the reasons they were trying things to make the work safer. You cannot work quickly or safely on wet steel.

In this trade, it's not about the tools; it's just your ability. The tools are basic ones a guy might carry: a spud wrench (that's an open-end wrench to fit a one-inch nut with a handle tapered to a point for stabbing bolt holes), a bull-pin, a sleever bar (it's about three feet long and used for prying) and an eight-pound beater for pounding the bull-pins or drift-pins into place. That's a pretty standard outfit even for an iron worker today. Now, I never carried a sleever bar. With my short legs, it would drag on the iron. But with four of us up in the air, there was always a couple of sleever bars with us. Sometimes you need it on the big iron. On the bridge, all the connections are tight. So we'd have to pry open the gusset plates or beat the shit outa it with an eight-pound sledgehammer. That's why we call it a beater. It's just raw power.

You had to believe you could do it. Otherwise you'd get laid off. Some guys just starting out—this was before they had an apprentice school—some of them could do it, and in three weeks, they were pretty good hands. They could handle themselves on the iron and do basic tasks without getting killed. Some never make it; they were too scared. I know one time, one guy almost fell on the Golden Gate. He just froze. It took three of us to break him loose and lower him down. He just froze.

When a guy freezes, he just grabs onto something and won't let go. It took three of us to get him loose. We had to tie him up and lower him down with a rope to a safe spot. He went home and never came back. But that happens to some guys, see, they just grab hold of something. I don't know what comes over them. That never happened to me. I could look down; it didn't bother me. People say, "Don't look down." It never bothered me. I would think about it though. I'd think, well, let's see, if I fall—this was on the Bay Bridge—if I fall I'll just double up. Roll up in a ball, try to hit on my back. I'll have a chance. I always thought I'd have a chance. I never had a fear of falling or slipping. I always figured I could grab on to something. You have to have that in you.

I don't know what or why—you have it inside you, or you don't. You just know you can do it. Now, if you didn't have it inside you, you couldn't do the work. You either got it, or you don't; you can't fake it. That's a lot of "*its*." I think the "it" is confidence. Without confidence, you'd either fall or get weak. See, to do this job you've got to have strength, you've got to have good hands, you've got to be sure-footed and you've got to be quick.

I was on the Marin side of the Marin tower. We were going from the tower north toward Sausalito. There was one raising gang going north and one gang going south toward San Francisco from both towers at the same time. That's called balanced erection. That's to keep the load-stress balanced on each side of the tower. So, I was working on that side, and we were about two-thirds of the way across. You see, the steel would come up from barges at the base of the towers. They'd raise it up and put it on flat cars and rolled it out to us. Then we would get it and put it in, you know, one or two bays at a time. Then we would roll the rig [crane] ahead, see, and start all over again—same thing: top cord, bottom cord, floor beams and stringers beams. Then the rivet gangs would follow along behind us.

The beams are about ten feet deep, you know, and about eighteen inches wide—widest thing up there besides the top and bottom cords. Well anyway, that's where it was. One Monday morning, I just started work. It was wet, cold and slippery, you know, from the fog. We were just getting ready to move the rig ahead for the day. So, I'd come over to help jack the stuff up to get the crane ready to move. So, as I was stepping around the rig I stepped onto one of the stringer beams. You've got to put your foot down straight. Well, evidently I didn't. I stepped out too far, and I slipped. As I was going down, I remember, I flipped three times, flailing around, you know? I didn't have any fear really, I'd thought about it before I fell and figured the net would catch me and I would bounce up and land on my feet like I'd seen on a trampoline in the circus. Well, I hit the net and the rocks at the same time. The net went down to the rocks. I came up. The first drop when I hit the rocks didn't hurt too much, just kinda knocked the wind outta me. Then I came back down—that's when I had the pain. But it didn't knock me out. Well, I was taken to the hospital right away. They put me on the flatcar, took me to the tower, put me in a basket and lowered me down to a boat.

So they took me to St. Luke's Hospital, and they put a body cast on me right away. I was in the hospital about twelve weeks. Later, I figured I'd fallen about forty-three feet. See, from the top of the deck to the bottom cord is about twenty-two feet. Below that was the net, which sagged all the way to the rocks. The Friday before, I'd gone on the net to pick up all the tools and things that had fallen into it. I noticed when I stepped out onto it, it would sag all the way to the ground. Anyhow, after that, after I got hurt, they took the net off the carriage and hung it from the bottom cord. I imagine the net was made of about three-eighths-inch line with four- to six-inch squares, like a fish net. It's tied to the bottom cord with three-quarters-inch line every fifteen or twenty feet or so.

A photograph from 1936 shows how close the net was to the ground near the Marin side of the bridge. Zampa fell from the top cord of the deck into the net near the Marin cliffs. This photograph was taken just days before Zampa's fall. *Author's collection.*

Anyhow, that's how I got to be a member of the "Half Way to Hell Club." I was the first one who fell into the net. There were eight of us who fell into the net those first few weeks. The first four got hurt. I was one of them. Within a week or so, other guys fell, and they were brought into St. Luke's Hospital. People came and took pictures of us shaking hands—jolly, you know. I remember a doctor telling me how lucky I was. "If I'm so lucky, what am I doing here?" I said.

We formed the Half Way to Hell Club right there in the hospital ward. There were four of us in the hospital at the same time. One guy said we should have us a club. Call it the "Gone to Hell Club" or some such thing. You see, whenever a man fell to his death, we said he'd "gone to hell" because no iron workers would be allowed into heaven. But I said, "No, we didn't fall to hell. We only got halfway." So that's why it's called the "Half Way to Hell Club."

Before I got out of the hospital, my buddies would come and visit me, and they'd tell me, "Al, you're all through. You lost your nerve. You'll

Seen here is the steel frame that carried the safety net. It was attached to the underside of the deck steel and could travel out ahead of the construction work. *Author's collection.*

never be an iron worker anymore." Now, this was 1936, and I worked all the way to 1970 before I retired. Well anyhow, I said, "I don't know," and I kept thinking about it. Then I said, "Ha, I know I can make it." So, when I got out of the hospital, instead of coming home, I went right out to the bridge and walked all over it. I was a little pale from being there, but everything was OK. It didn't bother me at all when I got ready to go back to work. See, I had a steel back brace on when I got out of the hospital. I didn't do that work for about two years. I was thirty at the time. After I got hurt, they took the net off the carriage and hung it from the bottom cord of iron so it wouldn't touch the ground. Why didn't they do that in the first place?

The first time they used a net was on the Golden Gate Bridge. We got spoiled on the Golden Gate Bridge. Guys would clown around on the net. You work a little faster with the thought of a net there below you. After the Golden Gate, most bridges used safety nets of some kind. On the second Carquinez Bridge, we had a net. When we were putting the steel up—you know, connecting—we had no nets. After we put the steel

Newspaper illustration of the scaffold that fell into the nets in February 1937. *Author's collection.*

up, then we put up the net for the guys coming behind us: the bolter ups, the welders and all that. See, so the net was there, but you can't pull it over before you've put the steel up. You can't pull the steel up through the nets.

Well, that net was a good thing. It saved my life. But it was also blamed for killing ten men. I was still in the hospital but heard about it right away. That was February 1937, and there were some guys underneath the road deck stripping the forms off. They were working on a scaffold below the road deck but above the nets. And, somehow or other, the work platform fell into the net. Thousands of pounds of debris, tools and men fell into the nets. If it would have went straight through, it wouldn't have been so bad.

I guess the net almost held. It wasn't made to support that much weight, and it just broke loose and went down. As each section went down it overloaded the next section and that gave way, too. It just kept ripping off and popping those three-quarter-inch manila lines. So the whole thing went crashing into the Bay. I think they lost ten men that day. They drowned, rolled up in that net like fish. It mostly got the carpenters and laborers who were stripping the forms for Pacific Bridge. No iron workers went in the hole that day.

There's a famous picture that appeared in one of the San Francisco newspapers that shows the net falling and five or six guys clinging to the net with little arrows pointing to the doomed men. Well, we come to find out

This oft-reproduced photograph that purports to show the men falling with the net. The men were added with ink to the empty net to create drama and sell papers. *Author's collection.*

later that the paper painted in those black dots with ink and said it was the men to sell papers.

So that was a bad wreck—not the worst that ever happened though. Back in 1907, they was building a bridge across the St. Lawrence River in Canada, and about midway through, the whole thing collapsed into the river, killing seventy-plus iron workers. Most of the dead were Mohawk Indians from the local reservation. This tragedy, I think, is where the idea that all iron workers are Indians started. People used to say to me how they thought all bridge men were Indians. "Sure," I'd say, "I'm a Cacciatore Indian."

Then in 1957, again in Canada, a bridge under construction collapsed, killing eighteen bridge men. That was in Vancouver at a place called Second Narrows. [That bridge is now called the Ironworkers Memorial Bridge.] Because of the name, "second narrows," people confuse that collapse with the famous collapse of the Tacoma Narrows Bridge up in Washington State. Two different bridges; two different wrecks.

I worked on top of the Golden Gate tower, seven hundred feet in the air. On the tower tops, we had a big derrick crane. Well, I was one of the guys who went up to help change the sheaves—that's the wheels at the tip of the boom. We would just climb up on the latticework of the crane boom, and the sheaves were big. They weighed fifty pounds or better, we had to take them out by hand and lower them down and bring the replacement back up by rope.

I always wanted to get better, learn a new trick from an old hand. I watched, listened and helped out wherever I could. As good as you are, there's always room for improvement. If you're spinning cables, they check the wire. They check the tension on the spinning wheel. No matter what we was doing, we never moved the floats until one of the inspectors came by to check our work. When he says OK and records in his notebook that he's satisfied, then we move. On the Golden Gate, they fabricated all the steel back East—in Indiana, I think. They transported it to a storage yard in Alameda. And before each piece was shipped out to the bridge, it was checked to be sure it would fit in place when we got it. But even then we had problems. Some of that iron was big. And it expands and contracts as the temperature changes during the day. So sometimes, we had to beat it into submission.

A lot of people said the Golden Gate could not be bridged. Nine out of ten would say, "No way, it can't be done." The water was so rough and the tide would go out so fast and the waves would be high. I worked over on the Oakland Bay Bridge before I came to the Golden Gate. It wasn't very rough there because it was sheltered from the ocean. It was on the Bay. But the Golden Gate is on the ocean—those ocean currents are a whole different story.

But we did it. And those of us that done it, we're still proud to be one of Strauss's boys. Strauss died just a year or so after the bridge opened. Maybe the strain of it killed him. Strauss, there was a man with both brains and

A statue of Joseph Strauss greets visitors to the Golden Gate Bridge. A statue of a riveter should also be at the plaza. *Author's collection.*

guts. But down there near the toll plaza, they have a sample of the main cable and a statue of Joseph Strauss. I think they should have a statue of a riveter. They really should. Something like that painting inside Coit Tower. One of those panels is a riveter.

Now, I never met Joseph Strauss, not like Maynard put in her play. I just seen him a few times. I never shook hands with him or nothing. I couldn't get to him; I was out there working. See, all the big shots (we call them all big shots), they were the ones that were doing all the hand shaking. In fact, they should have come out and shook hands with us. We were the ones doing the work. Of course, they designed it and proved every steel, every piece they designed, and they checked every piece of steel that went in there. Uh, they're very good. I'm sure glad they were.

But I saw him around a lot. We were busy working so we couldn't go shake his hand. That was for the upper men. Strauss was about five-foot-four, I guess—about my size. He went through a lot of crap to make the bridge a reality. He had brains. I'll tell you something about Mr. Strauss that most people don't know. One time, he was a skid row bum (at least that's the story we heard about him, but that coulda been a lot of "rags to riches" bullshit). A great man like him, brains and all. He was kicked around a bit. But he came back, and he designed and built that great bridge.

So, I'm most proud about my work on the Golden Gate Bridge. It's the more famous of the two. But the Bay Bridge is far more interesting. I worked on every part of the Bay Bridge. Until I went to the fiftieth anniversary celebration I'd never been across the Golden Gate. I never had no reason to go. I almost got killed there. But the Bay Bridge…When I go over the Bay Bridge, I can see myself up there. My fingerprints are all over that bridge.

Part 4

Back to the High Steel

I remember when they built the electrical towers between Crockett and Vallejo, across the strait, right next to where the new bridge is. I would walk up the hill after school sometimes and sit and watch them work. Then after work I'd walk home with Dad.
—Dick Zampa Sr.

When I went out to the bridge instead of coming home from the hospital, [my wife] knew I'd be going back because I told her it didn't bother me at all. "I'll go back as soon as I'm able." I had to wear a steel brace for two years, you know, take it off a little each day and put it back on. It kept my back straightened out. But Angelina left it up to me whether or not to go back to ironwork. She knew I was stubborn.

After the accident I was in no shape to go back to the bridge. I wore that back brace for two years. But I needed to support my family. My second son, Dick, was born in 1936 and I was getting $25.00 a week disability compensation from the bridge company. After about six months or so, it reduced it to $12.50 to encourage me to go back to work. I filed a claim with the Industrial Accident Commission because I felt the net was not properly installed. My suit kept getting postponed. After the bridge opened, all the guys I worked with, witnesses to the accident, they all scattered to the four winds. Taking jobs all around the country. So nothing came of my suit. If it happened today, I'd be a millionaire. My total compensation for that year was about $1,200, about half of the $11 dollars a day I earned on the bridge.

Well, I couldn't get by on the compensation. I needed a way to earn extra money, so I started renting boats out to local fisherman for a dollar a day. At one time, I had twenty-five boats that I would rent out. I kept them at Joseph's Fishing Resort in Rodeo. Well, somehow or other, the bridge company got wind of my side business, and they sent out a couple of guys to rent a boat. One of the guys offered me money to row them out so they can fish. Of course, they wanted a few pictures of the trip, and later the bridge company claimed I was fit to work. Now, I was still in a back brace. But the pictures made me look bad. So it was tough for a while. But we had friends in town and a good reputation. People would give us credit. A couple of times, we was three hundred dollars in debit. But when I worked a bit I'd pay them off. So, eventually, I had to find a real job. I took a job at the Mare Island Naval Shipyard, where I worked as a shop fitter for two years. I only took that job because I wanted to be close to home and spend time with my boys, Gene and Dick.

Later on, when my boys were in high school and they started to play sports (baseball and football), I was working as a rigger at the Union Oil for a contractor. Every Friday afternoon, the kids would play baseball, and I'd take off. The boss would come up and say, "Hey, you can't be taking off every Friday." I said, "What do you mean I can't." I said, "Listen, I want to tell you something. I've been raising my kids and waiting all these years to watch my kids play ball; they come first. Now, if you want me here, I take off on Fridays; otherwise, write it out and give me my pay and I'll go work someplace else."

That's the way you have to be in this kind of world. If you lay down, they will step all over you. Nobody steps on me—never! If I didn't like it, I would take off. If I liked it, I stayed. Sometimes it wasn't the work. It was the people I work with or for. If I didn't like them, off I'd go. I never had no trouble. I'd get a job any time (away from home, any time). I'd never starve. I had experience and was a good bridge man. I would get calls all over to go to different jobs. They all know all about me. "That guy, he's all right. Hire him, you'll make money with him. He works, and he knows what to do and how to do it."

I never worried about my ability. I had the confidence, and I knew that all I had to do is get the job. Once I got the job, I'm there because I work. I would be the last guy hired sometimes, and I'd be the last one there. Sometimes a guy might say to me: "This is the second or third job I've been with you. You're always in there. You come in later, and you're always in the end."

"Well," I said, "You ought to know. I see you working when there's work to do. You know what I know. Some guys just don't want to work. When they are paying you top money so you got to give them a good job." I always did, and I'd preach that to my kids, to the guys in union hall and everyplace else. I belonged to the union since 1934. I've been a union man ever since. I believe it's still the best.

Well, Angelina got used to my job. She always worried, but she accepted that I was iron worker. Then my sons wanted to do it. She said, "Why do you want to be an iron worker?" Gene said "Because dad travels around. He has adventures." I would go all over working. They would either call me or write me to come up to work. I'd go chasing rainbows—wherever they'd pay the most I'd go there. So, to a young man like Gene, that probably looked pretty good. But I left it up to them. "If you want to do it, do it," I said. "Sometimes it's good work. If you want to work with your hands I don't think you can beat it."

Back in the 1950s, I was on executive board for eight years. I helped run the Oakland Local 378. And then my sons both joined Oakland in '50s. Gene got in the iron workers first. Then my younger son, Dick, got into the trade about 1955, I think, after he graduated from high school in Crockett.

Then in 1955 or so, American Bridge came to town again to build a second Carquinez Bridge. I just looked at it and thought, "I don't know if I want to work there." I had other jobs paying the same money—easier. It's a little late in life to start risking it all on another bridge. At the time, I was working up in Fairfield, Travis Air Force Base, putting up these airplane hangars. On the way to Fairfield every morning, I'd cross the first Carquinez Bridge. So, after the Travis job, when I was between jobs, I'd head down to the Oakland 378 Union Hall on 1734 Campbell Street. Me; my two sons, Dick and Gene; and Jiggs Madden were sitting shooting the shit at Raimondi Park. It's across the street from the union hall, right near the Bay Bridge.

So, Arnold Pierce, the union agent, came out and talked to us. He said, "Al, they want fifteen men out there on the Carquinez Bridge. I can't get anybody to go out on that damn bridge. God darn, I'm worried."

You see the union is contractually obligated to supply labor to the contractors. If the union doesn't supply the man power, they can hire guys off the street. So, I said to Pierce, "What's a matter with the men they brought out with them?"

"Ah," he said, "they weren't making time, and they left." He said, "You guys have to go out to the bridge or they are going to hire guys from the street, non-union." It was raining a lot of the time, and a lot of the boomers followed American Bridge around from job to job. They're allowed to bring some of their key men, like I did earlier when I traveled to Texas for Missouri

Valley. But that year, it rained more than usual, and the men weren't getting enough hours. So they'd take off for other jobs. See, living on the road, we call that "booming out," and it can get expensive if you're supporting a family in another state. Pretty soon, there wasn't enough men, so they called the hall.

"Well," I said, "if guys off the street can do the work they're welcome to it."

"No, no," he said, "we got to keep control of our work. I know damn well if you go out, you and your boys, the other guys will follow."

So, I asked Gene and Jiggs, "How about it? If you guys want to go, I'll go; otherwise, if you guys say 'no,' then I won't go either." So Gene says, "Oh yeah, Dad, let's go. We'll do it. We live near there anyhow." They were young guys who wanted to test themselves against the bridge. All real iron workers want to work on a bridge. So I told Pierce, "OK," and we went out to the new bridge.

So, that's how me and the boys got out to that second Carquinez Bridge, and sure enough, the next week, here come all the rest of the guys. I guess Arnold Pierce said, "Well, if it is good enough for Al Zampa and his sons, it sure got to be good enough for you." So I helped bring them over there, me

By 1957, the second Carquinez Bridge was taking shape near Crockett. *Author's collection.*

and my two sons. See, a couple of guys had one son in the trade. But I'm the only guy that had two sons in the trade. And they were good. They had to be. And Arnold Pierce used that to his advantage. I explained to my boys many times what you had to do to be a good iron worker. Looking back at all the bridges—the Carquinez, the Bay Bridge, the Golden Gate, the Richmond–San Rafael Bridge—they were great projects. A lot of guys would give their right arm to say they'd worked on the Golden Gate. But to me, my biggest thrill was working on that second Carquinez Bridge with my two sons, Gene and Dick. That was very satisfying.

My life during the 1960s was pretty stable. I had plenty of work, and my kids were getting married and starting families of their own. In 1963 or so, people started getting interested in the Golden Gate Bridge and the Half Way to Hell Club. But not much came of it as I remember, certainly nothing like what happened for the fiftieth anniversary in 1987. I retired from the iron workers in 1970. I was sixty-five years old. On my last job, I was in the raising gang but on the ground, hooking on. I had the slings on the iron beam. I could read the blueprints. So I knew what order the iron goes up in. But on my last day, I told the foreman I wanted to go up in the air. Connect the iron. The foreman said, "If that's what you want, you can go up." And that's how I finished my career as an iron worker.

Well after I retired things were pretty quiet until the mid-1980s, when people started getting interested in the fiftieth anniversary of the Golden Gate Bridge. That's when my retirement years got crazy.

Part 5

THE MAKING OF THE LEGEND

DON ZAMPA (GRANDSON)

I think it was important to him to work as a connector on his last day. He wanted to prove that he could still connect and that he was retiring out of convenience rather than necessity.

DICK ZAMPA SR. (SON)

The Charles Kuralt interview for CBS Evening News is what really started the media train rolling for my father. Before that, Al's life was pretty quiet. Charles Kuralt had a series called *On the Road with Charles Kuralt* where he traveled around the country doing short documentary stories about interesting people and places. In 1987, Kuralt decided to profile some of the guys who built the Golden Gate Bridge. My dad was one of those guys. Once Al got on TV with Charles Kuralt, he got some notoriety. So when the anniversaries for the Bay Bridge and the Golden Gate Bridge rolled around people started calling to talk to Al. I don't even know all of the interviews he did. I'd come home from work and see news vans and cameramen and lights set up on his front porch. It got a little out of hand. Someone once told me they'd seen a story about him on TV in Germany. Then organizations started inviting him to make public appearances. Don, my son, drove him to

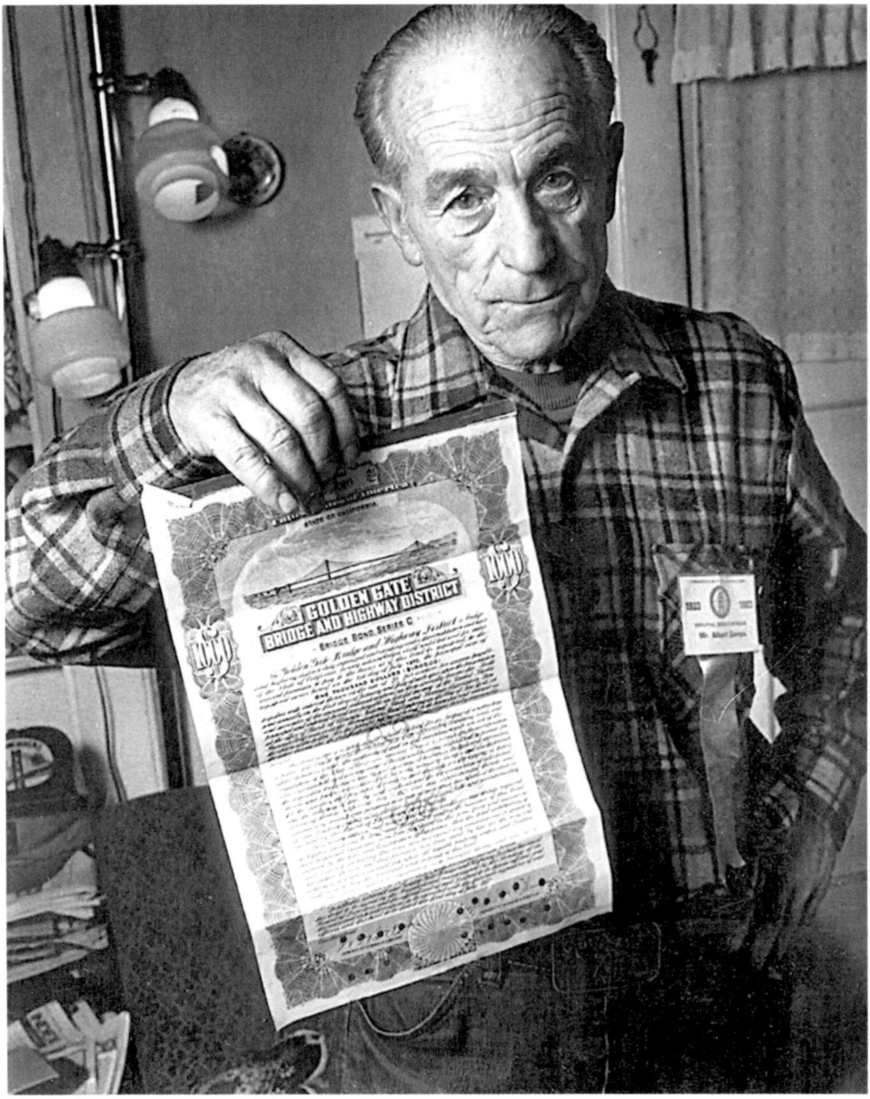

Zampa, holding an original $1,000 bond issued to finance the construction of the Golden Gate Bridge. It was presented to Zampa at one of the many celebratory functions he attended in 1987. *Courtesy Zampa family.*

a lot of those functions. Al didn't want drive to the city, so he might not have gone if not for Don. So, that combination of things got Al's name out to the public. He was a good talker and people liked him.

Isabelle Maynard (Author)

When I first met Al Zampa, I had been with the Tale Spinners Theatre group for about a year. The group was dedicated to collecting oral histories and turning them into dramas. Our mission statement read: "Tale Spinners creates and produces original theatre inspired by California oral histories. These histories are dramatized to illuminate their relevance to the lives of contemporary audiences. We pride ourselves on the diversity of plays we produce."

I joined the group intrigued by the idea of writing a play based on an oral history. Until that time, I had only written "from the heart"—poems, short stories and my memoir "China Dreams." The idea of researching a subject, going out to interview people and then dramatizing that material was a new challenge, one that I wanted to tackle.

In 1985, the group was given a long list of subjects compiled by the staff at Tale Spinners. We knew that 1987 was the fiftieth anniversary of the bridge's opening, and we wanted to do a play about the bridge. Some of the subjects that were made into plays that year were *Red Harvest*, about a suspected communist who was tarred and feathered in Sonoma County; *Mind's Eye*, about artists who painted the Coit Tower murals and the censorship they faced; *Boomtown*, about the Richmond Ship Yards during the war; and, of course, *The Ace* about the Golden Gate Bridge.

I went home that night with the bridge as my subject. Forty years earlier, I had sailed underneath that red span as a wide-eyed young immigrant from China. I'm not even Chinese, but that's a long story. (Read *China Dreams* to learn about that.) For people coming into California, the Golden Gate Bridge is what the Statue of Liberty is to people coming in from across the Atlantic—a symbol of the American dream.

But how do you write a play about a bridge? You have to have human beings to make a play. I went to the History Room at the San Francisco Library and started reading the newspaper stories about the bridge on microfilm. There were several newspapers in the city then, and they each ran regular articles about the bridge in 1936 and 1937. There were weekly reports by chief engineer Joseph Strauss, drawings of the bridge, union disputes and plans to celebrate the bridge's opening.

But it was a small item in the *San Francisco Chronicle* dated October 20, 1936, that caught my attention. Its title, "Net Grounded; Bridge Worker Falls; May Die," identified the injured man as "Al Zampa." I looked into the story and found that Zampa survived his fall into the nets. I realized that the way to

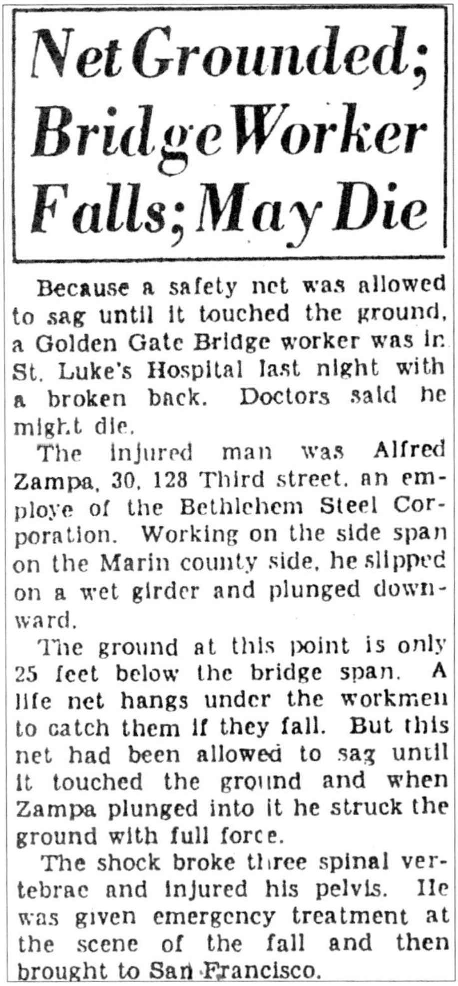

Net Grounded; Bridge Worker Falls; May Die

Because a safety net was allowed to sag until it touched the ground, a Golden Gate Bridge worker was in St. Luke's Hospital last night with a broken back. Doctors said he might die.

The injured man was Alfred Zampa, 30, 128 Third street, an employe of the Bethlehem Steel Corporation. Working on the side span on the Marin county side, he slipped on a wet girder and plunged downward.

The ground at this point is only 25 feet below the bridge span. A life net hangs under the workmen to catch them if they fall. But this net had been allowed to sag until it touched the ground and when Zampa plunged into it he struck the ground with full force.

The shock broke three spinal vertebrae and injured his pelvis. He was given emergency treatment at the scene of the fall and then brought to San Francisco.

This short news account from October 1936 reports the bare facts of Zampa's near fatal fall into the safety nets on the Golden Gate Bridge. This article started Maynard on her search for Zampa's story, which was made into the play *The Ace. Author's collection.*

go with the play was to reduce the bridge to one man's experience building it. I called several historical societies in the Bay Area, and someone suggested I call the Iron Workers Union. So, after many dead ends, I phoned Iron Workers Local 378 in Oakland and explained what I was doing and was surprised to learn that Al Zampa was very much alive, in his eighties and living in Tormey, a small village near Crockett. The Union man then gave me Al Zampa's phone number and encouraged me to call him.

When I met him and his wife, Angelina, in Tormey, I knew I had my subject. He was a very charming man and a bit of a flirt. He spoke like a poet. He spoke of "winds howling like banshees," of "scampering around the iron among the sounds of gulls," of wanting to "tame that devil bridge that tried to kill me." The vividness of his descriptions made me think of a trapeze artist. That's where I came up with the idea of using an acrobat to dramatize the physical nature of the work.

After interviewing him, I wrote the first draft of the play very quickly. Al's grandson Don Zampa, being an iron worker himself, was very helpful, too. That whole first part of the play, with the men talking on the bridge, Don helped write that. I mean, I don't know what iron workers talk about at work. Their talk gets very salty at times.

When I finished the play I realized I had taken some liberties with the story. For example, I put a scene in the play where Al meets Joseph Strauss.

After the premiere of *The Ace*, Al (wearing the suit) and Isabelle Maynard took questions from the audience and discussed Al's life and the writing process. *Courtesy Isabelle Maynard.*

That never really happened. So, since the play was a fictionalized account of Al's life, I decided to change the title character's name to "Al Bono." But most of the play is based on Al Zampa's oral history.

I didn't know what to expect that first night. I'm sure Al Zampa had never been to live theater before. I thought he might be a bit shy. I wasn't sure how he would react to the play itself. After the play, Mr. Zampa got up on stage with me to field questions from the audience. We were both moved by the attention and affection we received. *The Ace* is really a story about the resilience of the human spirit.

Archie Green (Labor Historian)

I first met Al Zampa on July 22, 1986, opening night of *The Ace*, an experimental drama based on his oral history. After the play, Zampa came out onto the stage with Maynard, the playwright, and some of the actors to talk about the play and his life. Zampa was a natural storyteller. He loved being the subject of a play. He wasn't surprised that a play was written about him. He just took it for granted that his experience on the bridge was interesting. He was the kind of worker that I've devoted myself to writing about. I've always been interested in the pride building tradesmen take in their work. Al Zampa was such a man.

Elma Garcia (Photographer)

I met Al Zampa in 1986 as part of the anniversary celebration for the Golden Gate Bridge. I had a photography studio in downtown San Francisco and was commissioned to take a formal portrait of Mr. Zampa. I called and made an appointment to see him one morning. His grandson drove him into the city. I didn't know what to expect before he arrived. I expected to see a frail old man, fifty years after the event and all. When he arrived, he was fit and walked with an air of confidence. I knew right away what I wanted to do. I had asked him to bring his hard hat and some tools. So I had him dress like he was going to work. While we were setting up he told me stories about working on the bridge. I wanted to capture a vintage feel, so I shot black-and-white film. I made fifteen or twenty exposures that day, and the photographs came out great. Everyone was very pleased. I don't think I ever saw him again after that. But when I lived in the Bay Area I would think about him every time I drove across the Golden Gate Bridge.

Dick Zampa Sr. (Son)

When the new bridge was breaking ground, we got a flood of invitations for Al to attend the ceremony. He almost didn't attend. He wasn't feeling well. His legs were going, and he needed a cane to get around. He was proud and didn't want to be seen as infirm.

I talked him into going, and once he got there, he perked up and was his old self. He made a little speech and told a few stories and everybody really liked it. Later, state senator John Burton called and said, "Why don't we try

On March 3, 2000, Al Zampa made his last public appearance at the groundbreaking ceremony of the third span across the Carquinez Strait. Al Zampa passed away just one month later, and shortly thereafter, the new bridge was named the Alfred Zampa Memorial Bridge in his honor. *Courtesy* Vallejo Times-Herald.

to get that bridge named after Al?" And it was John Burton who sponsored the legislation to get the bridge named for Al. As I recall, the resolution passed very quickly.

John Robinson (Author)

To have such an honor bestowed upon a Crockett native gave the new bridge even more meaning to the local community. A Bridge Celebration Committee was formed, and it soon planned numerous events for the opening celebration. The committee, chaired by Crockett's Gene Pedrotti, orchestrated a huge fireworks display off the new bridge on opening night. As part of the buildup to the opening festivities, the Bridge Celebration Committee held a logo contest offering $1,000 in prize money for the winning design. Local people were invited to vote for their favorite design, which

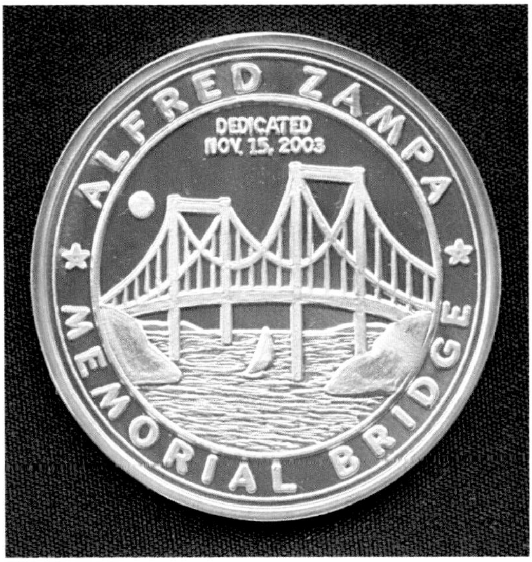

Left: The bridge logo as it appears on the commemorative coin produced by Gene Pedrotti and the Bridge Celebration Committee. *Courtesy Gene Pedrotti.*

Below: A view from February 2007 shows some of displays documenting Zampa's career featured at the Contra Costa Historical Society Center in Martinez, California. *Photograph by John V. Robinson.*

turned out to be the collaborative effort of Lori Cohee, Cleven Goudeau and Melissa Carlson-Loewe.

In late 2006, at a board meeting for the Contra Costa County Historical Society, we were discussing ideas for the quarterly exhibits at the History Center in Martinez. I suggested an exhibit about Al Zampa and the three

Along with photographs and historical bridge memorabilia, the museum exhibit contained a large metal replica of the Zampa Bridge, the sight of which in the window drew in many curious visitors off the street. *Photograph by John V. Robinson.*

Carquinez Bridges. At that time the second generation of great bridges were being built and people were interested in those projects. The Alfred Zampa Memorial Bridge opened in November 2003, the new Benicia-Martinez Bridge was nearing completion and the new Bay Bridge was half complete. Since Al Zampa was integral to building the early bridges, it was a perfect fit for our program, and the timing was right. We had a meeting with Dick Zampa Sr. from the Alfred Zampa Memorial Bridge Foundation, and with his help, we were able to get many of the exhibits for the display. He helped us locate an old rivet forge and the rivet-gang tools, as well as many hand tools from the Golden Gate Bridge period of bridge construction. They had a sixteen-foot model of the Al Zampa Bridge created for the opening day celebration. Local 378 had some of its apprentices come over one morning and cart all the exhibits into the museum. We had a nice storefront window on Main Street in Martinez. The model of the bridge was in the front window. It was a real eye-catcher. The exhibit ran from February through the end of April and was a great success.

Don Zampa (Grandson)

In 2005, I was business manager at Oakland Local 378 when I got a call telling me that some people in Ortucchio, Italy, were interested in the Alfred Zampa Memorial Bridge. I had no idea how they even knew about it. Maybe one of our relatives back there heard about the bridge somehow. Both of my grandparents' families came to America from Ortucchio, Italy. And as it turns out, several families in Crockett also have relatives or ancestors in that part of Italy. They sent a fax to the iron workers. As fortune would have it my father, Dick Sr., was president of the District Council, and I was at the local. So we were able to help them out. As I recall, the first few contacts were via fax machine. Then we progressed to phone calls. There are only a handful of people in Ortucchio who spoke English. I'd call them at 7:00 a.m. in morning, which is about 4:00 p.m. in the afternoon there, and tell them about Al and the bridge.

The Italians are proud of the fact that two bridges in the United States are named for Italians: the Verrazano Narrows Bridge in New York and the Zampa Bridge here in California. So they wanted to dedicate a piazza to Alfred Zampa, and they invited me to come over for the ceremony in September 2005. So I said yes, not knowing what to expect when I got on the plane to Rome. It was difficult because of the language barrier and the fact that I don't know them and they don't know me.

When I got there, I was met by a car and taken to a hotel. Ortucchio, a small town of about 1,500 people, is east of Rome. I arrived on a Wednesday, and the ceremony was not till Saturday. There was one woman in town, Claudia, who spoke English pretty well. I would have been lost without her help. She worked her ass off the whole time I was there. On the second day,

I was picked up at the hotel and taken to Ortucchio, where I was treated like a war hero or something. People cheered for me. I don't know who they thought I was. I went on a tour. I did interviews in the newspaper and RAI, the national radio and TV outlet, and I talked to school kids and was given a tour of an old castle. It was strange in a way.

And as it turns out, there were several people in that town who had relatives in Crockett. I recognized their names. So I was like, wow! This is interesting. It was like going back in time. The pace of life is slower there. One time we were driving somewhere and got stuck behind a guy and his ox pulling a cart. On the Saturday of the ceremony, there was a parade and a speech, and they gave me a key to the city. There was real happiness about having a native son return to the old country. I was asked to say a few words, and I got choked up a bit. They were so happy to have me as their guest of honor. I really didn't know how to react. I got emotional. Before I went over there, I had about two hundred postcards made up with the 1987 photo of Al that Elma Garcia took. I would give them out to the kids I'd meet. They loved them. I went over on a lark, but it turned out to be a very moving experience.

Part 6

THE ACE

By Isabelle Maynard

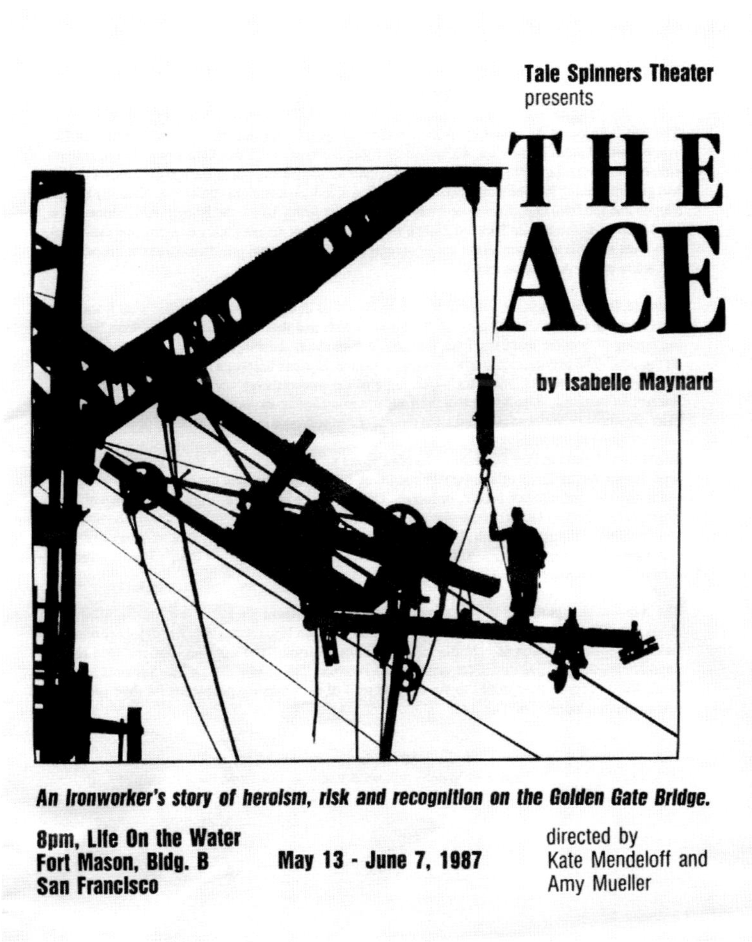

Al Zampa's exploits on the Golden Gate Bridge were the subject of a play written by Isabelle Maynard. Here is the cover of the program. *Courtesy Isabelle Maynard.*

Cast of Characters

Young Al Bono (in his thirties)	Iron Worker
Old Al Bono (in his seventies)	Iron Worker
Marie Bono (in her thirties)	Al's wife
Joseph Strauss (in his fifties)	Chief Engineer Golden Gate Bridge
Ernie (in his thirties)	Iron Worker
Rusty (in his thirties)	Iron Worker
Acrobat (wearing work clothes)	A symbol of strength and agility
Two Bridge Workers	
Radio Announcer	
Three Members of Press Corp	
Non-specific Voices (NSVs)	

Slides of building the golden gate bridge/introductory music.

Enter acrobat.

Lights on acrobat.

Worker sounds/bridge sounds.

Lights on workers.

Workers A and B begin to speak.

A: Boy what a day. San Francisco all fogged in.
B: Like pea soup.
A: Bolts out.
B: Damn. I've got a bad one. Gonna need a drift pin.
A: You asshole. Hit it like you live…hard!
B: Ease up.

Sound of hammer hitting steel three or four times.

A: Pin coming in.
B: OK, I've got it.
A: Rivet's in.

Sound of a rivet gun.

Young Al, offstage: You guys. We're going to jump the rig ahead.
A: One more rivet.
B: Man! They're right on our ass.
A: Want us to land them beams for you?
(Young Al, offstage): Yeah.
A: Signal the rig.

Workers A and B continue work activity.

B: OK [hand signals].
A: Beam comin' at you.
B: No problem.

Emergency whistle blows twice.

B: Goddamn, I think someone just fell.
A: Damn fog. Can't see. Jesus Christ!

Both workers freeze.

Lights on acrobat. Acrobat performs.

Lights out on acrobat.

Lights on Marie.

Marie: Oh my god!

Lights out on Marie.

Lights up on Joseph Strauss.

Strauss: God damn it!

Exit acrobat.

Lights on Old Al.

Old Al (present day): Al Bono. Iron Worker. Bridges have been my life. When I retired, I still was a connector. Up in the air like the man on the flying trapeze. A wonderful feeling comes over you when you're working bridges. They're beautiful…and scary. You gotta respect a bridge and not fool around. They used to call us sky-walkers. Bridge-monkeys. There we are climbing and running all over them cables. Scot-free. Looking down I seen the whole damn world spread out in front of me.

Enter acrobat.

Old Al: Steamers below looking like toy boats. Sometimes we'd be working above the fog, looking like little black dots in the sky. And all of us knowing all the time we was working on the greatest bridge of all times—one of the Wonders of the World. [*Music out*] The Golden Gate Bridge.

We didn't talk much about it, but we knew it. It bound us together like them cables binding the bridge. Bridge men are special folks. Men of steel. High-rise men got it easy—every third floor is covered. But we, mountain goats, we gotta have strength, fortitude and talent to climb up there in them skies. But what a bridge man needs most is confidence, spelled with a capital *C*.

A lotta people wouldn't even attempt bridge building. That's where the nerve and guts come in. Me, I've been attempting dangerous things ever since I was a kid. Like the time I swung on the elm tree in our backyard. Did it 'cause Eli Neugartner said it couldn't be done. He was the block bully. So I showed him.

Lights fade on Old Al.

NSVs
1: Come down from that tree, Al.
2: You'll fall. Dummy.
3: Boy, look at Al. He sure is high up.
4: Come down. Come down.
5: I'll race you across the hill, if you come down.

Lights up on Old Al.

Old Al: Sure was a good town to grow up in. Crockett. A company town that took real good care of us. Everything was free: gymnasiums, parks, Christmas parties, when all of us kids would line up and get big juicy oranges and shiny red apples. And when movies came in—well, every Friday, we'd have free movies in the gym. We had a wood stove to keep us warm and we raised our own vegetables. And folks was real friendly—if someone was building a house, folks would come by, grab a shovel and start digging.

Dad worked in the refinery. For lunch, he'd take a chunk of mother's bread, salami and cheese and a bottle of wine. Them was Prohibition days, and you wasn't allowed to take wine, except if you was Italian. Or Spanish. Anyone else they'd fire if they caught them with wine.

When Dad got lead poisoning, Ma said, "It's the only vacation your dad will ever have." He stayed in the hospital for a month, then went right back to work. He had five of us to feed.

We lived in the lowlands and them high mucky-mucks lived up on the hill. When us kids would cross town and start climbing the hill, the rich kids would yell at us, and if we made it to the top—the town watchman would run us back down.

NSVs
1: Dago, dago.
2: Dago, dago.
3: Scaredy cat.
4: Dago, dago.

Old Al: Dad said, "Pay no mind, Al. Pay no mind." But I hated it.

NSVs
1: Dago.
2: Dago.

Old Al: Got out of high school and worked as a meat cutter. Hated working inside, cooped up like a prisoner. Them was Depression days, and jobs were scarce. Sitting at lunch one day, a buddy said, "Hey, Al, you'd be a good bridge worker—you're tough, wiry and quick with your hands. And you're a daredevil. Hear there's a job opening up in Missouri." Well, Marie didn't like it none, but the pay was good, and I promised her she and the boys could visit if I stayed long enough on the job. She never did come, 'cause I kept moving. Always chasing rainbows, she said. I sure built me a lot of bridges—all kinds—suspension, camelbacks, cantilevers, railroad girders. In Missouri, Texas, Oklahoma and as far as Virginia. When I come back to California, Dad and Marie thought maybe I'd gotten bridges out of my system. "It's a job for desperate men, they don't care if they live or die crazy men," Dad said. But, when I was working the Bay Bridge, I'd watch them spinning cables on the Golden Gate, like silver threads when the sun shone on them, and I knew that's where I wanted to be.

So I joined Local 377 and proud to be working on the bridge that everyone said couldn't be done. I was the only native Californian—they come from all over the country to build that bridge. And they were good, too. A guy wouldn't travel three thousand miles if he didn't know he was good

Lights on workers.

Working sounds.

Enter acrobat.

Young Al and workers.
Ernie: Damn, they always push you this hard?
Rusty: Lately.
Al: Hi, Rusty.
Rusty: Hey, Al, meet Ernie here.
Al: Al Bono. You from around here?
Ernie: Nah, I'm from Missouri.
Al: Been there once. Beautiful country. Did ya ever work for Smitty at Missouri Valley Bridge & Iron?
Ernie: Nah. My old man worked for him last year.
Al: Hey, look. See? Ain't that Mr. Strauss down there?
Ernie: Can't really tell from here.

Strauss appears looking up at bridge.

Rusty: How long do you think it'll take this here apple to hit the water?
Al: I'd say a count of ten.
Rusty: Bet you it takes longer.
Ernie: Fifteen?
Al: (*throws apple*)
Ernie: One. Two. Three. Four…

Whoop! sound when apple falls.

Lights fade on workers.

Strauss leaves.

Music.

Lights on Old Al.

Old Al: One day, the wind blew my lunch bag over. I watched it gliding down, down, down like a brown bird. Sometimes the wind was raw, miserable. Damn winds coming all the way from Alaska, cutting though your clothes. Most of the time wasn't tied to nothing, but when it got too rough we had to use a number nine wire to hang on our suspenders and attach it to the rail.

Wind sound.

Guys get hallucinations up there. Start thinking—if I fall, what will I do. Imagine yourself doubling up, grabbing' on to something, clinging' to iron. I mean you're up there 740 feet on them tower tops. But you can't afford to think about slipping' or getting' hurt.

You kinda put your fear to the side and look up straight ahead. You hang on to your confidence, and you hang on tight.

Wind sound—acrobat scales rope ladder.

Seen a guy freeze up once. Took three of us to pry his hands from the steel—he was hanging on so tight. Me, I never feared looking down. No, sir.

Acrobat on rope ladder.

The scariest thing for me was them ferries we rode to the job. I was always kinda leery of them waves around me. And then there were the days it rained and we didn't work. We had good times in Sausalito.

Music.

Lights on acrobat.

Playing poker, shootin' dice and drinking. Pay was damn good. I was making fifty-five bucks a week. Chasin' women. Me, I had me a rainy-day gal or two. Nothin' serious, I just had to get my mind off the work sometimes.

But the real woman in my life, the one and the only, was Marie. My wife. I was crazy about her. You shoulda seen her when she was young. Green eyes. Wavy chestnut hair. A real beauty.

Lights on Marie and Young Al.

Marie: God, Al, I saw you falling off the bridge again. The same damn dream I keep getting.

Young Al: Hush. It was just a dream. Nothing else. You've got nothing to be afraid of. I'm here.

Marie: I don't think I can go through another one of those nightmares again. They keep coming, regular, since Buck slipped last month, I thought I would die.

Young Al: Buck was careless.

Marie: Louise said it scared Buck plenty but he won't admit it.

Young Al: He was singing and horsing around up there while we was rigging.

Marie: Singing?

Young Al: Yeah. You know Buck. Always out for a good time. It was one of them gorgeous days, and we could see straight to China. And Buck was singin' and having a good time. A gust of wind caught him, and he lost his balance. All he got was some bruises and a lot of sympathy from Louise. [*Pause*] Come on, Marie, let's get some sleep.

Marie: I can still see Mike Shauhnessy's face when he came to the door. I thought sure you was dead. "No, it ain't Al," he said. "It's Buck Horton. Al wants you to come to the hospital right away. To take care of Louise."

Young Al: Well, Buck's OK. You were there. He was in and out of that hospital in a couple of hours—they just patched him up and sent him home. After twelve years, I've never had an accident. 'Cause I'm real careful, and I'm the Ace. Remember that! So you got nothing to worry about.

Marie: What if it does happen to you? I'm getting so I'm afraid to open the door and see one of your buddies standing there.

Young Al: They built us a big beautiful net to protect us, remember? You should come out and see it—hangs there ready to catch us even if we fall. [*Pause*] It won't happen to me. I promise. Lucky is my middle name.

Marie: But what if it does happen to you? Why don't you quit, Al?

Young Al: Now you promised, Marie. You promised no more "quit" talk.

Marie: I keep hoping you'll change your mind. Now that you are a family man. I keep hoping you'll take up something that don't scare me to death. [*Pause*] I seen you, Al, hanging by your feet, head down. I seen blood all over your face, and I seen fire all around you. I see it in my dreams. Getting to be so I'm afraid of going to bed.

Young Al: You gotta stop talking that way, Marie. You're going to make me lose my nerve if you keep that talk up.

Marie: I want you to lose your nerve, Al.

Young Al: What'd you say?

Marie: Maybe that will put some sense into you. What am I going to tell the boys when they grow up? Be like your dad, boys. Start building bridges and scare your wives to death every time you leave the front door!

Young Al: Any more of this talk, I'm going out to take a walk. A long walk.

Marie: Go on. Leave. Don't listen to the truth. You're chicken, Al. Can't take the truth.

Young Al: I'm going.

Marie: Go take a walk. Or maybe go visit your rainy-day girl.

Young Al: Cut that out. I don't wanna hear that kind of talk.

Marie: You and me, Al, we got whole lists of things not to talk about. Can't talk about falling off bridges, can't talk about your rainy-day girl—or girls. Can't talk about Buck. What can we talk about?

Young Al: The good things in life. The picnic next week. The tomatoes I'm planting. The money we're saving to go on vacation. And your pretty eyes. And your lovely hair.

Marie: What about that other girl?

Young Al: Come on, honey. You know you're the one and only.

Marie: Do I? How?

Al crosses to start dancing with Marie.

Marie: Oh, Al, you'll be the death of me yet.

Young Al: I'm wild about you! I'm gonna kill you with my loving.

Carries her to bed.

Acrobat enters.

Marie: Guess I'm stuck with you. My trapeze artist. Me worrying about you. My heart pounding every time you leave that door. And you swinging up there in the blue skies, having a great time. I guess there is no curing you. [*Pause*] I oughta leave you, Al Bono. That's what my mother says. She says I should before I'm left a widow with two little ones to bring up.

Young Al: Now, gal, you ain't going nowhere and you know it. Don't listen to your ma. You'll feel better tomorrow. It's Sunday, and after church, we'll take the boys to the park. [*Holds her*] Your Al is here. Tell the boys their dad is the man on the flying trapeze.

Marie: Don't you get dizzy up there?

Young Al: The only time I get dizzy is when I look at you. [*Kisses her*]

Dancer on trapeze.

Music.

Lights off Al and Marie.

Lights on Old Al.

Old Al: Marie never understood what a thrill it was—stuck with me, though. Them was rough times when we was building the great bridge. Depression days. Unemployment all over. Soup lines. Families moving in together and sharing rent. Some people living in the streets. I remember looking down from the tower and seeing hundreds of men below. They was cooking bacon and beans in five gallon tins, and they was looking up at us. I could almost smell their cooking. They stood around all day to see if one of us would quit, get fired…or fall.

[*Pause*]

We was damn glad to have the job. Proud to do forty hours a week, come home knowing you were doing a job that damn few people had the guts to do. Then there were them engineers—they was something else, too. Inspecting everything. Every piece of steel, every rivet, every cable. Wanted everything meticulous, working with feeler gauges in one-thousandth of an inch. And the safety net they built us sure was something. Costs a lot of bucks anyway—$125,000.

The chief engineer, Mr. Strauss—a small man, but a mighty thinker. Heard he had trouble in the city—people didn't trust him. Disagreed with him.

Lights on Strauss.

Lights on Three Reporters.

Reporter A: Mr. Strauss. Would you please comment on the condition of the bed rock? And on the wind tolerance of the bridge in a storm? The tides here are the most treacherous in the world.

Reporter C: There has been great concern in the city Mr. Strauss that ships will not be able to go under the bridge and that planes flying in the fog will fly into the towers. What is being done about this?

Reporter A: We have heard that in an earthquake, the whole Bay will be littered with debris from the fallen bridge, making the Bay impassable. Would you care to comment, Mr. Strauss?

Reporter B: Mr. Strauss, people are saying that this bridge is sucking the tax dollar from the tax payer. Do you really think the people of San Francisco need this bridge?

Strauss: Ladies and gentlemen of the press, our world of today revolves around things which at one time couldn't be done, because they were supposedly beyond the limits of human endeavor. I am not afraid to dream. And I believe in hard work and an absolute unwillingness to be discouraged. My bridge will stand as a monument to man's highest achievement. It will be Statue of Liberty of the West, the gateway to the western world.

Reporter C: Mr. Strauss, the men could get dizzy spells up there and could fall.

Reporter A: We are also told that your safety net is not foolproof and could tear under great weight.

Reporter B: Even now the net is not taut.

Reporter C: Mr. Strauss, I have heard that engineers estimate there will be one death per every million dollars of construction cost! Is that true, Mr. Strauss?

Strauss: My safety record speaks for itself. We have doctors at Fort Point planning diets for the workers to control their dizziness. The workers are ordered to wear hard hats newly designed by my staff. The net is the most up to date in the business, and the Golden Gate Bridge is the first bridge to have this incredible safety device.

Reporter A: Mr. Strauss, has it ever been tested?

Strauss: We are in the midst of a depression. Let us not forget the Golden Gate is the only good thing happening in the whole country. It will be a single span being built under conditions that everyone claimed could not be done. A highway in the sky; an incredible road of shining metal, spanning cliff to cliff.

Reporters A, B AND C. [*Cacophony of all three voices*]: Mr. Strauss, Mr. Strauss, Mr. Strauss.

Strauss leaves followed by reporters.

Lights on Old Al.

Old Al: One Monday morning, I just started work. The fog was real thick. Couldn't see further than my hand. They were pushing us real hard. We was getting ready to move the rig ahead. In order to do that, we had to move the support beams. I come over to help out. Done it a thousand times before: top cord, bottom cord, floor beams, then the stringers. Something distracted me, maybe a gull flying low.

Acrobat enters and performs in background.

I put my foot on the stringer. It's quite a deep step, and you gotta put your feet down straight. Real straight. I stepped out too far. Lost my balance. It was wet, slippery from the fog. I hit my heel and fell. [*Pause*]

Wind howls—acrobat falls.

Going down flipped three times and rolled around. Had no fear though. I had thought about what would happen if I fell. I figured the net would catch me. I remember thinking I'd land on my feet like on a trampoline. Like a clown you see in the circus. But I hit the net and the rocks at the same time.

Boy did it hurt. Pains shooting down my left leg like electricity. Couldn't feel nothing down my right leg. I remember thinking—maybe this is it. Finito. Maybe one of my legs was maimed for good—like an insect squashed by a club. At least I wasn't going out with the tide. [*Pause*]

They lowered me down in a basket and took me to the hospital right away.

Knock on door.

Lights on Marie.

Marie: Oh, my God. Not Al. Please not Al.

Lights out on Marie.

Lights on Strauss who stands silent.

Lights out on Strauss.

NSV: Extra, extra, read all about it. Net grounded. Bridge worker falls. A safety net touched the rocks, Golden Gate bridge worker, Al Bono is in St. Mary's Hospital with a broken back. May die. Read all about it. Extra. Extra.

Lights on Old Al.

Old Al: They put me in a body cast, and I stayed in it for three goddamned months.

Lights on Young Al in hospital bed. Knock on door. Ernie enters.

Ernie: Sorry to wake you up. Been watching you sleep. You been snoring.
Al: I never snore.
Ernie: Well, I was standing listening to you outside the door, and I heard you sawin' away, old buddy.
Al: Aw, cut it out. [*Pause*] Good to see ya. Gets kinda lonely around here.
Ernie: Let you out of the body cast?

Al: Three damn months is long enough.

Ernie: Yeah. So. How they treating you? All right?

Al: You bet. Nothing but the best for the Ace. You should see Nurse Riley, she's really a dish…whoo-ee! Gives me rubdowns all the time. Washes me in bed.

Ernie: Yahoo. Not bad. Not bad at all. I should be so lucky. [*Pause*] Is that today's paper?

Al: No. Yesterday's.

Ernie: DiMaggio got another hit yesterday.

Al: That Dago is a hell of a ballplayer. Hated to see him go to the Yankees.

Ernie [*laughs*]: Yeah.

Al: Did Rusty take you again?

Ernie: Yeah.

Al: Can't wait to get out to a ballgame with you guys. The Oaks!

Ernie: The Seals! We'll go. Soon. You'll be out on your feet in no time.

Al: Sure.

Ernie: No time at all. Yep. You'll be out there pounding the pavement…yeah.

Marie enters.

Marie: Hi, hon. Hi, Ernie. You look good today.

Al: Hey, looks like Rusty took Ernie again. Got my paper, hon? I wanna check the box scores. Gotta keep up with these guys.

Marie: I forgot.

Al: Well, how about going downstairs and getting me one? What's the matter, cat got your tongue?

Marie: (Looks at Ernie) Al….

Ernie: Al, I guess I better give it to ya straight. Rusty fell yesterday. It's all over the front page of the *Chronicle*.

Al: We're gonna have us a regular club here at St. Mary's. What room is he in?

Ernie: He ain't in no room, Al…Rusty went in the hole.

Al: What happened?

Ernie: The stripper ripped from its hangers.

Al: Christ!

Ernie: Corner brackets tore loose. I always thought if ten tons of steel hit that net it would rip it to shreds. They kept telling us the braces were safe. Goddamn.

Al: So what happened?

Ernie: Well, I was up above and heard something that sounded like a giant piece of cloth ripping apart, something being torn. I thought maybe it was an earthquake. I knew Rusty was with the gang on the stripper underneath. Then I heard a sharp crack.

Al: Go on.

Ernie: I heard screams and things falling and sounds like thunder when the stripper ripped from its hangers. Then everything seemed to stand still, and all I could hear was my heart pounding. I started praying, Al. I did. Then I looked down, and there was Rusty all fouled up in the net, with all that wreckage and beams pinning him. And him struggling to get free. There were other guys falling, going down in the net, but all I could see was Rusty. [*Pause*] I must have gone crazy, Al. I spread myself on this beam and reached down. I guess I thought I could save him. I could see Rusty's face and…it looked all white and he was looking up at me. I knew Rusty was gonna die. I just knew it, but I didn't feel nothing, Al. Nothing at all.

Al: Yeah.

Ernie: The other guys. Dick Rader. Gus. Clarence. Shorty Dumbarton. He just turned eighteen.

Al: Damn kid.

Ernie: Yeah. Just a kid. Listen, Al. I been thinking. I saw Jake at the hall…and I talked to him…about you…I mean there's a real good chance…that you can get an inside…

Al: Inside work? Forget it, Ern. I'm an outside man. A bridge builder.

Ernie: You were a bridge builder, Al. But you fell…

AL: What're you telling me?

Ernie: I'm telling you, old buddy, that anyone who falls off that bridge can't ever get up the nerve to get back on it. You know that.

Al: What the hell are you talking about? I'm the Ace. Remember?

Ernie: You'll always be the Ace. You're the best all-round man, Al, on that bridge. No one's gonna take that away from you. But you fell. You broke your back.

Al: I'm gonna get back on that bridge, Ernie. I'm not gonna let that devil bridge get me.

Ernie: Wake up, Al. You ain't gonna be able to climb them cables no more. A bridge man's gotta have good hands and good feet. Remember? You're the one that always was saying that.

Al: I mend fast, Ernie. [*Pause*] I just gotta get out and prove that I still got the nerve.

Ernie: How long do you think you'll last up there on the bridge? Hobbling around. Clinging to the rail. You're not going to be able to look down, Al. It's a long way down. A long way. I want you to think about that, Al.

Al: What do you want me to do? Peddle shoestrings. You've said your piece. Now leave me alone.

Ernie: You gotta quit, Al. Think about it. Think real hard. 'Cause I tell you, buddy, I am. We lost Rusty. Well, I'll leave you alone for a while. Want anything next time I come?

Al: How 'bout a beer?
Ernie: You got it. I'll bring you a whole case.
Marie: Seems like I'm always mending. Boys just keep growing out of things. Guess they're gonna need new clothes. Thought I'd go downtown and do some shopping. Those boys are real good, Al. You'd be proud of them. Last night they helped me clean out the basement and didn't make any fuss at all about doing it. They've been a real help.
Al: I'd like to be alone. Really. Don't worry. Just go home and fix the boys a nice dinner. Go. See ya tomorrow.

Marie leaves.

Dancer enters—climbs ladder.

Al: Rusty. Redheaded Rusty. I got him into the union.

Crack—crash—music.

Acrobat falls.

Music fades. Sounds of waves crashing.

I gotta see if I still got the nerve. I gotta.

Acrobat performs.

Lights out.

Scene in semi-darkness. Strauss standing at bridge rail.

Young Al enters smoking, whistling "Marie."

Strauss: I would greatly appreciate it if you would not whistle.
Young Al: Sure pal.
Strauss: Thank you.

Young Al starts whistling and quickly stops himself.

Young Al: This is my first time at night. Sure is pretty. Used to work on them towers.
Strauss: Used to?

Young Al: Yep, before the accident. Maybe you've read the papers about me. How I fell. And lived. None of them docs thought I would.

Strauss: I wish you the best in your recovery.

Young Al: Hey, ain't you Mr. Strauss, chief engineer?

Strauss: Yes, yes, I am.

Young Al: What are you doing out here at night? I seen you from up there during the day, but I never figured you come at night. But I guess it's your bridge so you gotta right to come any time you wanna.

Strauss: I often come here to survey my work. Take stock. At a time when no one can recognize me. I like being alone.

Young Al: Sorry if I intruded. Guess I think of it as my bridge, too. My fingerprints is all over the south tower. I love this bridge.

Strauss: You must excuse me. I really did not mean to be rude, Mr.…?

Young Al: Bono, Al Bono.

Strauss: I do appreciate you men who have been working on…my bridge for so long.

Young Al: It's all right, Mr. Strauss. I can see how you might feel kinda special about this here bridge. Kinda like it's your baby that you've nursed for years.

Strauss: I, Mr. Bono, have been thinking, living and breathing this bridge for twelve years. We're inseparable.

Young Al: I've been living and breathing bridges for years, too, and the Golden Gate for three years. Ain't nothing that needs to be done on a bridge, I can't do. Rigging, riveting, connecting. At the union they call me the Ace.

Strauss: The Ace…I, too, know all there is to know about bridges.

Young Al: Ain't that something? Cigarette?

Strauss: Tell, me, Mr. Bono, why are you here in the middle of the night?

Young Al: I come to see if I still have the nerve.

Strauss: The nerve?

Young Al: Yeah. See my buddies tell me I'll never work on the bridge again 'cause they think I've lost my nerve. After Rusty…Rusty Donovan…went down the hole…I…Well…So I come to see if I can still climb on that devil bridge.

Strauss: Can you?

Young Al: Dunno. Standing here talking to you it feels fine. But before, I was standing by myself, and I looked down and thought I was hallucinating. Down there, in the water, I seen my face and around it a ring of fire, like I was burning all over. Got dizzy and had to look away.

Strauss: Well, I certainly understand how you would consider giving up bridge work, Mr. Bono, after your experience.

Young Al: Let me ask you, Mr. Strauss. What would you do if you had fallen off?

Strauss: Probably same as you. Probably would come and face it and see if I could do it. Conquer it. Tame it.
Young Al: Never thought you and me—we'd be thinking alike. Funny, ain't it?
Strauss: I often thought Napoleon and I…well, that we were alike…I mean, we both were small men…we both persevered and never gave up. When I was a child and spent many hours in bed with asthma, I came across a picture of Napoleon on the Bridge of Arcole. He looked so majestically alone. And so sure of himself. He became my hero. And I never forgot the sight of that silver bridge, glittering behind him.

Young Al winces.

Strauss: Are you in pain, Mr. Bono?
Young Al: Nah. Just a spasm down my back. It got broken you know.
Strauss: Yes, I know. We spent weeks figuring out the net. The best net in the world. The first bridge ever built with a net. Costs $82,000, Mr. Bono, and $42,000 to install. Nothing but the best for my men. What went wrong?
Young Al: It's OK, Mr. Strauss. Accidents happen. Nothing is guaranteed when you're working on bridges.
Strauss: I promised, Mr. Bono. I promised the investors. I promised the people of San Francisco. I promised myself there would be no more accidents. No accidents, Mr. Bono. Do you hear me? No accidents.
Young Al: I guess you did your very best. Your engineers, they examined every damned rivet on this bridge. Can't blame them.
Strauss: How can you be so charitable?
Young Al: I just wanna forget the whole thing and go on about my business. No sense brooding. But, sometimes I'm afraid I won't be able to forget it, and then I think my nerve is gone and that means I ain't the Ace no more. So who in the hell am I?
Strauss: I told you I come here to survey my work. I lied. I come here to find courage to go on. To see if I have the flame in my heart to persevere, to build monuments. Do you know how many bridges I've built?
Young Al: Nope.
Strauss: Four hundred. All over the world. In Leningrad, Canada, Mexico, Belgium. All over the U.S. Many men have died working on my bridges.
Young Al: How many?
Strauss: I've lost count. I used to know. I've stopped counting.
Young Al: All jobs are dangerous. Gotta friend who's a meat cutter. Lost three fingers butchering a hog.
Strauss: Mr. Bono, I have been responsible for the death of many men. Good men. Like Napoleon. Building bridges is like a military campaign. You sacrifice lives to win. The question is—is it worth it?

Young Al: Is it?

Strauss: I don't know. I come here to find out. At night. I come here to listen.

Young Al: What do you hear, Mr. Strauss?

Strauss: The humming of steel. The crashing of waves. I hear the wind whistling through the cables. And the excitement…the danger…of building bridges grips me. I begin the dream of the Parthenon. I see the sculptured splendor of the monument of Victor Emanuel II in Rome. I see the Cathedral in Milan, the Hermitage in Leningrad, the Taj Mahal. And so on and so forth.

Young Al: You been to all them places?

Strauss: Some.

Young Al: Well, I've been around the U.S. of A. building all kinds of bridges. But I must say the Golden Gate is the mightiest, right out in the ocean. Not for all guys of course—some just couldn't take it…I guess I'm wondering now if I can. I've been trying to shake the picture of myself falling…but it keeps coming back to me. I gotta stop thinking about it. Gotta get my nerve back.

Strauss: If you listen to your fear, you'll join the ranks of the fearful. Fear and doubt will immobilize you, Mr. Bono. It will kill you.

Young Al: Ain't you ever thought of quitting?

Strauss: Every day now.

Young Al: What keeps you going?

Strauss: This bridge, perhaps one or two others. When my life is extinguished, my bridges will remain. That comforts me, Mr. Bono. Besides I'm damn good at building bridges—shame to waste all that talent.

Young Al: Same here. I don't wanna waste my talent.

Strauss: So how are you going to prove it?

Young Al: Prove what?

Strauss: That you have the nerve, Mr. Bono. That's what you came to find out, isn't it?

Young Al: I got it.

Strauss: Yes?

Young Al: I'll sit on the rail. Stretch my hands out. Not hold on to anything. How about joining me?

Strauss: Me?

AL: Hell, no one will see you. We'll both show her we can tame her.

Strauss: Her?

Al: She excites me like a woman. How about it, Mr. Strauss?

Strauss: Maybe another time. You go ahead, though.

Al [*gets on rail*]: Damn! I still got the nerve!

Acrobat enters. Music.

Lights out on Young Al, Strauss and acrobat.

Lights on Old Al.

Old Al: After that meeting with Mr. Strauss I thought to myself that maybe I was the only one in town who knew the real chief engineer. I didn't tell no one about it. Not even Marie. He never asked me not to, but I felt it was a secret between us, and I was real proud of that. We had become almost buddies, him and me, both of us fighting and conquering our fears.

While I was still out of work, I got $25.00 a week for the first eight months and $12.50 a week for the next eight months. Wasn't really enough to get by on, so I got me some boats and took people out fishing in the bay. Caught us some great big bass. Kept me busy while I was figuring out about going back to bridge building.

Tried suing them on the negligence of the net. I never got a dime. Rusty's family sued, too, but none of us got nothing. Today, I'd probably be a millionaire, the way people are suing and getting. But somehow it don't feel right to me—getting a million bucks for a job you love.

People was good to us. Gave us credit, knew the Bono family had fallen on bad times. I'd buy $100 worth of provisions and the folks at the Oakland Produce Company would say, "Don't worry, Al, we know you. You'll pay later, we have faith in you."

After I fell, a couple more guys fell into the net. Didn't hurt nothing, just got bruised. Guys use to yell when they fell, "Well, here we go. Looks like a nice long buggy ride." We formed a club called Half Way to Hell Club, and the newspapers come out and took pictures of us.

So, the Bono family was doing all right, all things considered.

Lights down on Old Al.

Marie enters. Lights on Marie.

Al enters.

Music.

Al: Think I'm going to sit this one out. But you keep dancing.

Marie: No, honey. I've got a lot of work to do.
Young Al: For me!
[*Marie dances*]
Remember them days we used to go dancing at the Avalon? Before the boys came?
Radio Voice: And you will be able to hear Victor Young and his orchestra at the spectacular Golden Gate celebration coming up next week at Crissy Fields.
Al: I tell you what, hon, you and me, we'll be dancing together when they celebrate the bridge's fiftieth. You'll see.
Marie: Don't make promises you can't keep. People are talking on and on about the festival. My boss at the Alta Mira asked me if I wanted to participate. Join the parade. They're having some sort of a float that the Alta Mira is sponsoring.
Al: Well, what did you tell him?
Marie: I told him no.
Al: Why'd you do that? Could be fun. Maybe me and the boys could get a free ticket.
Marie: I told him no because I wasn't going to celebrate that bridge that almost took my husband's life.
Al: It's also been our livelihood. Good money. Remember?
Marie: Al, you're thinking of going back to being a bridge worker, aren't you?
Al: Yeah.
Marie: For heaven's sake, why?
Al: I've told you a million times. I am an iron worker. A bridge man. It's my life, and a man don't give up his way of life. Cause there ain't nothing I can't do on a bridge. And cause there's no thrill like it. You oughta forget the accident, honey, and look ahead. But you're like an elephant. Never forget nothing.
Marie: You almost got killed there, Al. That ain't nothing. Seems like there ain't a thing ever bothers you. You drive me crazy with your forgiving and forgetting nature.
Al: Father Mulcaney should hear you. He's always preaching about forgiveness.
Marie: That's for saints. And I'm no saint. I hate that bridge, and I hate them celebrating it. All I can think of is your accident and Rusty's death.
Al: Stop hanging on to bad thoughts, Marie. It eats you up.
Marie: How can I stop? You're still limping, and it's been a year since the accident. I hear you at night, Al. Your back is aching and you moan. You don't hear yourself, but I hear you. You just won't admit it.
Al: I'm getting better, honey. They're paying me compensation while I'm recuperating.
Marie: Pennies.

Al: What about the fishing boat, Marie?

Marie: Even with that and my waitressing we're not getting by.

Al: We're doing fine, Marie. I think we have lots to celebrate.

Marie: Al Bono! Sometimes you drive me up a wall. We gotta live on credit, and you saying we got something to celebrate.

Al: I'm alive ain't I? So life is tough, but I'm still the Ace.

Marie: So what did you get being an Ace. Two inches in the papers.

Al: You gotta pay for your love.

Marie: What are you takin about?

Al: I love the bridge. I love working on it. So sacrifices have to be made. Risks taken. Like Napoleon. He never gave up.

Marie: What are you talking about? What's it got to do with Napoleon?

Al: Nothing, Marie. Just something I heard one day.

Marie: Who do you talk about Napoleon with? One of your drinking buddies? Or maybe that rainy-day gal.

Al: Cut it out, Marie. It was just something I heard. Long time ago.

Marie: You are a funny one, Al. Still surprise me. Anyway, I told the boss no thank you, I will not go around parading on no float.

Al: Suit yourself, gal. I know you got a mind of your own.

Marie: Sometimes I think I get so mad because I have to be mad for the both of us.

Al: No use getting mad, I say. Live and let live.

Marie: That drives me bananas. Here's all this celebrating being planned and nothing in it for you. For us. I hate being left out. Forgotten.

Al: You just don't like other folks having fun it you're not.

Marie: It's not fair.

Al: Who said life was going to be fair?

Marie: There you go again. Give them the other cheek. Boy! Father Mulcaney sure would be proud of you. But I don't believe none of that stuff. That meek shall inherit the earth is not for me. I am a fighter.

Al: Damn it, Marie. So am I. I fought every time I went up on that bridge. I fought the fear, and I've conquered it. And I did it by myself—without talking and yapping about it all the time. I've still got my confidence and my pride. What else do I need?

Marie: Something special. For falling off that damn bridge.

Al: Like what?

Marie: A letter from the governor. A seat in the parade. Money.

Al: You'd probably like a whole float called "Celebration to Al Bono who fell off the bridge!"

Marie: That wouldn't be bad. Or a float dedicated to the men who died. Or a lifelong pass to cross the bridge. A pass for the whole family, including your mom and dad.

Al: You're a dreamer, Marie. A pass would be nice, though. Maybe I'll get one someday.

Marie: Hell will freeze first, Al. Mark my words. You ain't going to get nothing. And I ain't going near that celebration. They'll have to celebrate without Marie Bono.

Lights fade—Marie exits.

Lights up on Young Al.

Radio Voice: And now the floats are turning into Crissy Fields. The Wild West show is a real beaut. Cowboys and cowgirls. Some of the prettiest young people in San Francisco. And there—someone is milking a real cow! This is unbelievable! And right behind them is the First Battalion, and all the way from our Canadian neighbors in Vancouver, here to help us celebrate. And what a celebration this is. Yesterday at least three thousand people participated in this Span of Gold pageant—all whom have spent weeks preparing for this great day. There goes the Slovenian Alliance float with girls in native costumes and the maypole.

Ladies and gentlemen, this is the moment we have been waiting for. The crowd is beside itself. They're waving American flags. Children are being hoisted on their parent's shoulders.

Our chief engineer Joseph B. Strauss flanked by city officials is heading toward the reviewing stand, where he will give his dedication speech. Ladies and gentlemen, Mr. Joseph B. Strauss.

Lights on Strauss.

Strauss: Ladies and gentlemen, my heart is full. The mighty task is done. The impossible dream realized. In one leap, San Francisco has emerged from bridge obscurity to bridge supremacy. San Francisco, whose hub is the center of the Golden Gate, has chosen life. Tonight we pause to celebrate that phenomenon in the birth of a bridge. We have fastened a cut-steel necklace around the throat of San Francisco and festooned it with long white ribbons of roadway and sidewalks. This bridge was launched amidst a thousand hopes and fears, damned by a thousand hostile sneers. Yet never its course was stayed. Ladies and gentlemen, victory is ours.

[*Clapping and cheering*]

Perhaps nothing can better express my feelings at this moment than a few lines from my poem "Victory" written especially for this occasion. If you will indulge me.

I am the thing that men denied.
The right to be, the urge to live.
And I am that that men defied.
Yet I ask naught for what I give.

Above the fogs of scorn and doubt.
Triumphant gleams my web of steel.
Still shall I ride the wild storms out
And still the thrill of conquest feel.

Lights out on Strauss.

Lights on Old Al.

Old Al: Soon as I could, I told Jake Townsend at Local 377 that I was ready and able to go back to work. Went on to build other bridges, the Carquinez, the American, the Amarillo and other bridges.

The sons followed me. And even the grandson. We're the only family in the whole USA that has three generations of officers in the same local at the same time. I've been a union man for fifty-two years. The Golden Gate Bridge was the first 100 percent union job. Damn proud of that.

Since I retired I miss working bridges. Hearing that wind blow, feeling it on your face. The feel of iron under your fingers. Not tied to nothing. Your courage deep inside you like a tight ball. Up there 740 feet high. Close to heaven and close to hell.

Them folks who drive across bridges—they just go across. Pay the toll, whiz by and never stop to look up or look around.

You know, I've never gone over the Golden Gate. Never. Why should I? I almost got killed there. I never had no reason to go. Oh, sure I've cursed that devil bridge. The lives it took. The damn cold. But I sure loved it, flying with the birds, touching the sky. Sometimes I go near it. Drive around in my car, get close and look up. I see myself all over that bridge. [*Pointing*] I was there and there and there.

My fingerprints are all over that bridge. You can't erase that, because that bridge is mine.

Exit Old Al.

Lights on Young Al and Workers A and B.

Young Al: Boom down. Easy. High! Got it.
B: Coming down. High! Got it? It's yours.
A: Up a hair. Little more, little more. High! That's good.
B: Let's go. Cut her loose. OK, you got her.
A: [*Whistles twice*]
Young Al: Let's get outta the wind.
A: [*Whistles*]

Exit all.

Curtain.

Pictured here are Al Zampa (left) and one of his fictional alter-egos who narrates the story. *Courtesy Isabelle Maynard.*

Appendix 1
Senate Concurrent Resolution No. 97

Resolution Chapter 135

Senate Concurrent Resolution No. 97—Relative to the Alfred Zampa Memorial Bridge. [Filed with Secretary of State September 12, 2000.]

Legislative Counsel's Digest SCR 97, Burton. Alfred Zampa Memorial Bridge

This measure would designate the westbound span of the Carquinez Bridge as the Alfred Zampa Memorial Bridge in honor and recognition of Alfred "Al" Zampa. The measure also would request the Department of Transportation to determine the cost for appropriate signs showing this special designation and, upon receiving donations from nonstate sources covering that cost, to erect those signs.

WHEREAS, Alfred "Al" Zampa was born on March 12, 1905, in Selby, California, the oldest son of southern Italian immigrants, and the brother to two brothers and two sisters; and

WHEREAS, After graduating from high school, Al Zampa went into business and became the owner of a meat market in Crockett, California, until about 1924, when a customer asked him if he wanted to go to work for that customer on the bridge they were building from Crockett to Vallejo; and

WHEREAS, When after some hesitation, Al Zampa decided to give it a try; and

WHEREAS, The first Carquinez Bridge opened in May of 1927, in part due to Al Zampa's efforts, and that bridge was to be the first of many bridges Al Zampa would work on in his illustrious career as an iron worker; and

WHEREAS, Al Zampa continued working with the company that built the Carquinez Bridge and worked on projects and bridges in Stockton, California, and later in Arizona and Texas, returning to California in the early 1930's to work on the Oakland–San Francisco Bay Bridge and the Golden Gate Bridge; and

WHEREAS, On October 20, 1936, this outstanding iron worker fell into the safety net while working on the Golden Gate Bridge and broke four vertebrae in his back; and

WHEREAS, During his years of recuperation following that accident, Al Zampa and a couple of his friends founded the area's first baseball league for the local youth; and

WHEREAS, He later returned to iron work and worked on the second Carquinez Bridge in the 1950's with his two sons, Richard L. (Dick) and Gene; and

WHEREAS, Al Zampa also worked on the Martinez Bridge and the Richmond–San Rafael Bridge and continued to work as a respected iron worker until he retired at the age of 65; and

WHEREAS, Many years after his retirement, and prior to the Golden Gate Bridge's 50-Year Celebration, Al Zampa received much deserved recognition and notoriety; and

WHEREAS, In 1987, he was the subject of a stage play entitled "The Ace" that was performed at Fort Mason in San Francisco, was interviewed by news stations from all over the world, and was interviewed by Charles Kuralt for the national television show "On the Road"; and

WHEREAS, Al Zampa was also interviewed for the History Channel on top of the building of the Golden Gate Bridge and more recently for a new show entitled "Suicide Missions: Skywalkers" which depicts the history of the Iron Worker Union; and

WHEREAS, Al Zampa was always proud to be a 65-year-old union member and grateful to have received a union pension after retiring from Iron Workers Local 378, Oakland, in 1970; and

WHEREAS, Al Zampa strongly believed in supporting the labor endorsed candidates, from the days of Franklin D. Roosevelt to Bill Clinton, and was a staunch lifelong Democrat, having never missed an election; and

WHEREAS, Al Zampa passed away on April 23, 2000, at the age of 95, and will be missed by many; and

WHEREAS, It would be a fitting tribute to Alfred "Al" Zampa to name the soon-to-be-completed westbound span of the Carquinez Bridge as the Alfred Zampa Memorial Bridge; now, therefore, be it

Resolved by the Senate of the State of California, the Assembly thereof concurring, That the Legislature hereby designates the westbound span of the Carquinez Bridge the Alfred Zampa Memorial Bridge in honor and recognition of Alfred "Al" Zampa; and be it further

Resolved, That the Department of Transportation is requested to determine the cost of appropriate plaques and markers, consistent with the signing requirements for the state highway system, showing this special designation and, upon receiving donations from nonstate sources covering that cost, to erect those plaques and markers; and be it further

Resolved, That the Secretary of the Senate transmit copies of this resolution to the Director of Transportation and to the author for distribution.

Appendix II

Glossary of Iron Worker Terms

ace: (n) a top iron worker.
asshole: (n) a bad kink in a wire rope.
beater: (n) four- or eight-pound sledgehammer. See also: Monday, single-jack and double-jack.
boomer: (n) term for men who travel out of town looking for work.
boom out: (v) to go out of town for a job.
bull pin: (n) tapered pin about twelve-inches long, used to align holes for rivets or bolts. It has a mushroom head for striking with a sledgehammer.
coon: (v) technique used to climb beams and columns using hands and feet like a raccoon.
drag: (n) money advance before the usual payday.
drag-up: (v) to quit a job.
endo: interjection yelled to stop the forward or backward motion of a piece of steel.
eyeball it: (v) to check alignment of something using the eyes only.
float: (n) a makeshift scaffold you hang from a beam. It's a four- by six-foot panel of boards with four thick rope lines in each corner to hold it up. Riveters used it to work from.
go in the hole: (v) to fall. See also: the Hole.
headache: (n) interjection used to alert someone of falling debris.
headache ball: (n) heavy round ball attached to the crane's load line. See also: pill.
hell dog: (n) pneumatic rivet buster.

APPENDIX II

This view from April 2004 shows the three bridges shortly before the deconstruction of the 1927 bridge was started. *Photograph by John V. Robinson.*

high: (adv.) interjection yelled to stop raising the steel. Short for "high-enough."
the hole: (n) any space that a person can fall into.
horse-cock: (n) large hand tool used by rivet gang.
Joe McGee: (n) generic name for anyone, or anything, that is not good enough.
Monday: (n) eight-pound sledgehammer. See also: beater.
pile butts: (n) a term used for pile drivers.
pill: (n) heavy, round ball attached to the crane's load line.
punk: (n) an apprentice iron worker; (v) to carry something.
pusher: (n) term used to describe demanding foreman.
raising gang: (n) the gang of men who hook the steel to the crane and set the steel in place in the air.
red ass: (n) to get very angry.
rig: (n) generic term for a crane, derrick or gantry, used to lift the steel into place.
ringer: (n) term used to describe a forty-hour workweek.
sand hog: (n) term for men who dig tunnels and work underground with compressed air.

A view from the Crockett shore showing the Zampa Bridge and the 1957 bridge after the demolition of the 1927 bridge that once stood in the middle. *Photograph by John V. Robinson.*

spud: (n) a temporary fastener, large or small, that is easy to insert and remove. Temporary piles are called "spud piles" by pile drivers.
spud wrench: (n) wrench with a tapered handle that can be used to help align rivet or bolt holes.
topping out: (n) to place the highest piece of steel in a structure, usually topped with a ceremonial flag and a small Christmas tree.
whistle-bit: (v) to still be working or putting tools away at quitting time.
wobble the job: (v) to take informal actions that disrupt the job site briefly.
wobblies: (n) slang for Industrial Workers of the World, also known as IWW.
yo-yo: (n) slang term for a pneumatic impact wrench used to tighten bolts.

Appendix III

Annotated List of Bridge Books

I have been interested in bridge history and have collected many interesting bridge books over the years. Listed here are some of the best for those readers who would like to know more about this subject.

Adams, Charles F. *Heroes of the Golden Gate.* Palo Alto, CA: Pacific Books, 1987.

Annan, Jason, and Pamela Gabriel. *The Great Cooper River Bridge.* Columbia: University of South Carolina Press, 2002. This book documents the building of the Cooper River Bridge that opened in 1929.

Bartelme, Tony, and Jessica Vangeren. *The Bridge Builders and Charleston's Grand New Span.* Charleston, SC: Evening Post Publishing, 2005. This book documents the construction of the new Cooper River Bridge.

California Department of Transportation. *Spanning the Carquinez Strait.* Oakland, CA: CalTrans, 2003. Greg Bayol, Bart Ney and Brian Boal produced this book about the Zampa Bridge that features the photography of Bill Hall.

Cassady, Stephen. *Spanning the Gate.* Mill Valley, CA: Squarebooks, 1979. This book contains fine documentation of the construction of the Golden Gate Bridge.

D'Acres, Lilia, and Donald Luxton. *Lions Gate.* Burnaby, BC: Talonbooks, 2005. This work documents the building of the Lions Gate Bridge in Vancouver, Canada, in 1939.

Gotchy, Joe. *Bridging the Narrows.* N.p.: privately published, 1990. This account of the building of the Tacoma Narrows Bridge is told from a worker's perspective. It includes many nice vintage photographs.

Haw, Richard. *The Brooklyn Bridge: A Cultural History*. New Brunswick, NJ: Rutgers University Press, 2005. In this cultural history of the most famous bridge in the world, Haw looks at the bridge from every conceivable angle.

Hobbs, Richard. *Catastrophe to Triumph: Bridges of the Tacoma Narrows*. Pullman: Washington State University Press, 2005. This scholarly account of the first two bridges across the Tacoma Narrows in Washington State contains a great collection of historic photographs.

Jamieson, Eric. *Tragedy at Second Narrows: The Story of the Ironworkers Memorial Bridge*. Madeira Park, BC: Harbour Publishing, 2008. This book is about the 1958 collapse of the Vancouver, Canada bridge that killed eighteen workers when it collapsed while under construction. That bridge is now called the Ironworkers Memorial Bridge.

Mallard, Henri. *Building the Sydney Harbour Bridge: The Photography of Henri Mallard*. Melbourne, AUS: Sun Books, 1976. This hard-to-find book features Henri Mallard's photographs of the great bridge from circa 1931.

Moore, David. *To Build a Bridge, Glebe Island, Sydney Australia*. Neutral Bay, NSW: Chapter & Verse, 1996. Photographer David Moore documents the construction of the Glebe Island Bridge in Sydney, Australia.

Rastorfer, Darl. *Six Bridges: The Legacy of Othmar H. Ammann*. New Haven, CT: Yale University Press, 2000. This work is about O. Ammann, the designer of six iconic New York bridges.

Richman, Steven. *The Bridges of New Jersey*. New Brunswick, NJ: Rutgers University Press, 2005. This book is a nice photographic survey of the many beautiful bridges of the Garden State.

Robinson, John V. *Al Zampa and the Bay Area Bridges*. Charleston, SC: Arcadia Publishing, 2005. This book is a photographic look at some of the lesser-known bridges of the San Francisco Bay Area.

———. *Bridging the Tacoma Narrows*. Crockett, CA: Carquinez Press, 2007. This work is a photographic look at the third suspension span across the Tacoma Narrows in Washington that opened in 2007.

———. *Building the Benicia-Martinez Bridge*. Crockett, CA: Carquinez Press, 2007. This work is a photographic account of the newest bridge between Martinez and Benicia in California that opened in 2007.

———. *Spanning the Strait: Building the Alfred Zampa Memorial Bridge*. Crockett, CA: Carquinez Press, 2004. This book is a photographic look at the construction of the Alfred Zampa Memorial Bridge between Crockett and Vallejo in California that opened in 2003.

Rockland, Michael Aaron. *The George Washington Bridge: Poetry in Steel*. New Brunswick, NJ: Rutgers University Press, 2008. This account is a cultural history of the George Washington Bridge between New York and New Jersey.

Schwartz, Harvey. *Building the Golden Gate Bridge: A Workers' Oral History*. Seattle: University of Washington Press, 2015. This beautifully edited book tells the story of the great bridge from the workers' point of view.

Stackpole, Peter. *The Bridge Builders: Photographs and Documents of the Raising of the San Francisco Bay Bridge, 1934–1936*. Corte Madera, CA: Pomegranate Artbooks, 1984. Written by a young Peter Stackpole, this book contains photographs of the building of the Bay Bridge from 1934 to 1936.

Talese, Gay. *The Bridge*. Photographs by Bruce Davidson. Illustrations by Lillian Rethi. New York: Harper & Row, 1964. This work tells the story of the building of the Verrazano-Narrows Bridge.

Trachtenberg, Alan. *Brooklyn Bridge: Fact and Symbol*. Chicago: University of Chicago, 1965. This work is a good synthesis of the Brooklyn Bridge as it is represented in history, literature and art.

Wortman, Sharon Wood. *The Portland Bridge Book*. Portland, OR: Urban Adventure Press, 2006. This book contains a detailed survey of the many and beautiful bridges in and around Portland, Oregon.

Bibliography

Adams, Charles F. *Heroes of the Golden Gate*. Palo Alto, CA: Pacific Books, 1987.

Baum, Willa K. *Transcribing and Editing Oral History*. Nashville, TN: American Association for State and Local History, 1991.

Berson, Misha. "Fingerprints on Steel." *San Francisco Examiner: Image*, May 10, 1987.

Blakely, Scott. "He Built Them All, Large and Small." *San Francisco Chronicle*, November 4, 1986.

Cassady, Stephen. *Spanning the Gate: The Golden Gate Bridge*. Mill Valley, CA: Squarebooks, 1979.

Daily Review (Hayward, CA). "Ex-construction Workers Swap Bridge Stories." January 5, 1983.

Dillon, Richard, et al. *High Steel: Building the Bridges Across the San Francisco Bay*. Berkeley, CA: Celestial Arts, 1979.

Gladysz, Thomas. "Pride and Danger in Building a Bridge." *West County Times*, May 20, 1987, B4.

Kuralt, Charles. "The Bridge Builders." *On the Road with Charles Kuralt*. New York: Putnam, 1985, 94–101.

Lochner, Tom. "Famed Bridge Builder Al Zampa Dies." *West County Times*, April 25, 2000.

McGinty, Brian. "Span of Gold." *American West*, April 1987.

Pereira, Joseph. "Bridge Builder Lived to Tell the Tale." *West County Times*, September 22, 1986.

Raddue, Gorden. "How He Joined the Half Way to Hell Club." *Contra Costa Independent*, January 6, 1983.

Rego, Nilda. "Gambled Lives on This Job." *West County Times*, November 12, 1986.

Schwartz, Harvey. *Building the Golden Gate Bridge: A Workers' Oral History.* Seattle: University of Washington Press. 2015

Seaver, Linda. "Bay Bridge Was Builders' Baby, Despite the Danger." *Oakland Tribune*, October 26, 1986.

Spencer, Richard. "Carquinez Bridge Is Granddaddy of Them All. *West County Times*, October 18, 1986.

Van Der Zee, John. *The Gate.* New York: Simon and Schuster, 1986.

Wells, Peggy Sue. "Carquinez Bridge Spans Sixty Years." *Benicia Portfolio*, May 1987.

Zampa, Al. Interview with author. Tormey, CA, August 1995 and August and September 1998.

Index

About the Authors

Isabelle Maynard (1929–2007) was born in Tientsin, China, and immigrated to California in 1948. An actor, writer, painter and oral historian, Maynard is best known for her book *China Dreams: Growing Up Jewish in Tientsin* (University of Iowa Press, 1996). She also published dozens of short stories in literary magazines and is the author of the play *The Ace: An Iron Worker's Story of Heroism, Risk, and Recognition on the Golden Gate Bridge*, produced in 1986.

John V. Robinson is an award-winning photographer and writer who is the author of six previous books, including *Spanning the Strait: Building the Alfred Memorial Bridge* (2003) and *Bridging the Tacoma Narrows* (2007). He holds degrees from UC Berkeley and San Francisco State University. In 2006, Robinson was awarded a Guggenheim Fellowship in the field of folklore and popular culture. In 2007, he was a California Council for the Humanities recipient for documentary photography. John Robinson teaches English at the California State University–East Bay in Hayward, California.

By Kyle L. Robinson.